Technofeminist Storiographies

Communicating Gender

Series Editors: Diana Bartelli Carlin, Saint Louis University; Nichola D. Gutgold, Pennsylvania State University; Theodore F. Sheckels, Randolph-Macon College

Communicating Gender features original research examining the role gender plays in communication. It encompasses a wide variety of approaches and methodologies to explore theoretically relevant topics pertaining to the interrelation of gender and communication both in the United States and worldwide. This series examines gender issues broadly, ranging from masculine hegemony and gender issues in political culture to media portrayals of women and men and the work/life balance.

Recent Titles in This Series

Adolescence, Girlhood, and Media Migration: US Teens' Use of Social Media to Negotiate Offline Struggles, by Aimee Rickman
Consuming Agency and Desire in Romance: Stories of Love, Laughter, and Empowerment, by Jenni M. Simon
Michelle Obama: First Lady, American Rhetor, edited by Elizabeth Natalle and Jenni M. Simon
Women in the Academy: Learning From Our Diverse Career Pathway, edited by Nichola D. Gutgold and Angela R. Linse
Communication and the Work-Life Balancing Act: Intersections across Identities, Genders, and Cultures, edited by Elizabeth Fish Hatfield
The Global Status of Women and Girls: A Multidisciplinary Approach, edited by Lori Underwood and Dawn Hutchinson
Technofeminist Storiographies: Women, Information Technology, and Cultural Representation, by Kristine L. Blair

Technofeminist Storiographies

Women, Information Technology, and Cultural Representation

By Kristine L. Blair

LEXINGTON BOOKS
Lanham • Boulder • New York • London

Published by Lexington Books
An imprint of The Rowman & Littlefield Publishing Group, Inc.
4501 Forbes Boulevard, Suite 200, Lanham, Maryland 20706
www.rowman.com

6 Tinworth Street, London SE11 5AL, United Kingdom

Copyright © 2019 by The Rowman & Littlefield Publishing Group, Inc.

All rights reserved. No part of this book may be reproduced in any form or by any electronic or mechanical means, including information storage and retrieval systems, without written permission from the publisher, except by a reviewer who may quote passages in a review.

British Library Cataloguing in Publication Information Available

Library of Congress Cataloging-in-Publication Data Available

ISBN 978-1-4985-9303-8 (cloth : alk. paper)
ISBN 978-1-4985-9304-5 (electronic)
ISBN 978-1-4985-9305-2 (pbk. : alk. paper)

∞™ The paper used in this publication meets the minimum requirements of American National Standard for Information Sciences Permanence of Paper for Printed Library Materials, ANSI/NISO Z39.48-1992.

Printed in the United States of America

Contents

Acknowledgments	vii
Introduction: From His Story to Their Stories	ix
1 Parallel Lives and the Recovery of Women in the (His)Story of Computing	1
Interrogating Cultural Assumptions about Women and Technology	5
Recovering Ada Lovelace and Hedy Lamarr	15
Conclusion: Re-Telling the Story	22
2 Distinguishing Rhetoric from Reality in Early Computing Culture	25
The Perception of Women's Roles: Now and Then	27
Dramatizing Early Computer Culture	31
Women's Programmed/Programming Lives	33
Grace Hopper: Queen or Pirate	37
Conclusion: Giving Credit Where Credit is Due	44
3 Bridging the Technological Gender Gap On and Off the Screen	53
The Steve Jobs Mythos	53
The Social Network	62
The Challenges of Leaning In	64
Silicon Valley: Art Imitates Life?	67
Conclusion: Making the Problem Visible	71
4 Gender Play and the Marketing of Misogyny	75
Still Trying to "Disrupt the Pink Aisle"	77
Gender Trouble in Silicon Valley	79
Technofeminist Remix Pedagogies	83
The Shared Search for Our Inner Wonder Woman	86
Creating a Counter Narrative	88

	One Step Forward, Two Steps Back?	93
	Conclusion: Confronting the Princess Problem	94
5	Sustaining a Technofeminist Future for Women and Girls	99
	No Time Like the Present?	101
	Technofeminist Re-Codes	105
	Technofeminist Back-Talk	108
	Coda	111
Bibliography		117
Index		131
About the Author		137

Acknowledgments

I am extremely grateful to Lexington Books Acquisitions Editor Nicolette Amstutz for her unfailing support of this project and her amazing patience and flexibility. Without her, *Technofeminist Storiographies* simply would not be published. I am grateful as well to Assistant Editor Jessica Tepper for her phenomenal assistance in formatting, and to Assistant Editor Jessica Thwaite in helping to keep the many components of this book organized. An equal amount of appreciation is due to those who provided anonymous feedback on the project; these recommendations were immensely helpful in expanding the book's audience and theoretical frame.

I have been fortunate in my career to work with many technofeminists who have helped sustain my own research interests. Specifically, I would be remiss not to acknowledge Dr. Jen Almjeld, who was a continued source of inspiration as she sent countless resources and encouraged me to stay on deadline. She is my colleague, friend, and frequent collaborator, as are Dr. Christine Tulley and Dr. Angela Haas, both of whom have been significant influences on my professional life. Other important collaborations influencing this project include those with Dr. Estee Beck and Dr. Mariana Gronowski. I also am indebted to Dr. Radhika Gajjala, who talked me through the process of writing this book and frequently reminded me, "You can do this!" She is my technofeminist role model. Other role models include Professor Emerita Gail Hawisher and Professor Emerita Cynthia Selfe, two feminist mentors who have guided my career over two decades. I shall be ever grateful to Cindy for our numerous coffee conversations in her living room in Columbus, Ohio, helping me formulate the ideas for this project in early 2015.

I also thank my husband, Kevin Williams; his love and encouragement has meant more than anything to me; he is what gives my life joy. And I must thank my mother, Angela Blair, who, as I like to tell people, ensured that for every toy I received, I also received a book. That balance contributed greatly to the development of this project. I am fortunate to have her in my life.

I am grateful to Routledge Press for the ability to reprint portions of chapter 1, which originally appeared as "Technofeminist Storiographies: Talking Back to Gendered Rhetorics of Technology" in *The Routledge Handbook of Digital Writing and Rhetoric*. Finally, I am also grateful to the Smithsonian Archives Center and to Stewart Lamb Cromar of the University of Edinburgh for permission to reprint the images that appear in chapters 2 and 4 respectively.

Introduction

From His Story to Their Stories

In a 2017 Center for American Progress Report "The Women's Leadership Gap: Women's Leadership by the Numbers," Judith Warner and Danielle Corley summarize a series of disturbing statistics involving working women in the United States: that despite being a gender majority at 50.8 percent of the American population, and earning the majority of bachelor's and master's degrees, along with nearly half of all law and medical degrees and specialized master's, these equal or majority data sets drop drastically when examining women's leadership roles. Referring to this dilemma as a "stalled revolution," the report outlines the pervasive gender and race gaps in executive and management roles across a range of industries: from the academic, to the legislature, to the entertainment community, and to the information technology industry. Citing data from the U.S. Equal Opportunity Commission, the report notes that "in 2014 women were just 20 percent of executives, senior officers, and management in U.S. High Tech Industries. As recently as 2016, "43 percent of the 150 highest earning public companies in Silicon Valley had no female executives at all." The data for women of color across these industries are far worse, with leadership percentages in the single digits. In comparison to other countries, for while the United States ranks first in educational attainment, it "ranks 26th in women's economic participation and 73rd in in women's political empowerment." Based on these data sets, the report concludes that parity with men is unlikely until roughly 2085, some 67 years away, or two to three generations from now. Similarly, the May 2016 Equal Employment Opportunity Commission Report "Diversity in High Tech" focuses on two challenges facing gender and ethnic diversity: "the 'pipeline' problem—STEM occupations attracting white men—and the inhospitable culture in relevant industries and occupations, forcing women and minorities to tolerate the environment or leave."

My rationale for introducing *Technofeminist Storiographies: Women, Information Technology, and Cultural Representation* with this labor data includes an effort to contextualize from an ideological, political, and economic standpoint the ways in which women's histories of technology have been situated in larger cultural ecologies that both shape and are shaped by assumptions about women's role in society. Based on these and other data sets cited throughout this book, parity appears an elusive goal for women working and studying in STEM-based fields. Moreover, technofeminist Judy Wajcman (2010) notes that "Feminists have pointed out that the problem [of excluding women from technoscience] does not lie with women (their socialization, their aspirations, and values) and that we need to address the broader questions of whether and in what way technoscience and its institutions can be reshaped to accommodate women" (p. 145).

Nevertheless, in her canonical book *Gender Trouble: Feminism and the Subversion of Identity*, Judith Butler (1990) outlines the problem of focusing on women as the subject of feminism in the ongoing "assumption that the term women denotes a common identity" (p. 4). Butler further argues:

> The political assumption that there must be a universal basis for feminism, one which must be found in an identity assumed to exist cross-culturally ... often accompanies the notion that the oppression of women has some singular form discernible in the universal of hegemonic structure of patriarchy or masculine domination. The notion of a universal patriarchy has been widely criticized ... for its failure to account for the workings of gender oppression in the concrete culture contexts in which it exists. (p. 3)

Technofeminist Storiographies addresses both Butler's and Wajcman's emphasis on context by foregrounding the historical and contemporary narratives of technological innovation that enabled and constrained women's participation within them, with a specific focus on computer culture. Butler's performative theory of gender has been adopted to "disrupt the categories of the body, sex, gender, and sexuality and occasion their subversive resignification" (p. xxxi) in ways that mesh with a primarily third wave feminist focus on intersectional theories that moved away from essentialized experiences of what were mostly white, middle class, heterosexual women to instead privilege more diverse race, class, gender, and sexual identities within and across cultures. As Patricia Hill Collins and Sirma Bilge (2016) outline in their introductory chapter to *Intersectionality: Key Concepts*:

> When it comes to social inequality, people's lives and the organizational of power in a given society are better understood as being shaped not by a single axis of social division, be it race or gender or class, but by many axes that

work together and influence each other. Intersectionality as an analytic tool gives people better access to the complexity of the world and of themselves. (p. 2)

While crediting Kimberlé Crenshaw (1989) with coining the phrase "intersectionalism," Collins and Bilge stress that the goals of this framework have deeper roots: "Intersectionality's core ideas of social inequality, power, relationality . . . complexity, and social justice formed within the context of social movements that faced the crises of their times, primarily the challenges of colonialism, racism, militarism, and capitalist exploitation." Intersectionality resists and rewrites "authoritative histories" that "represent partial views of the world," which are typically white and male in the United States, and in the case of this book, have an impact on the evolution of computer culture as we know it today.

To avoid essentializing women as a universal group and better address the need for the cultural complexity for which Crenshaw, Collins and Bilge, and others have called, this book relies upon a technofeminist storytelling methodology to recover both historical and contemporary stories of women's lived experiences with technology and within technological industries. I refer to this as a "storiography," a re-writing and re-telling of women's technobiographies to disrupt and talk back to larger cultural narratives that have excluded their voices and contributions. An early influence on my own work has been Cheris Kramarae, whose 1988 collection *Technology and Women's Voices* chronicled the impact of a range of daily technologies—the telephone, the washing machine, the microwave—on women's lives, concluding that such technologies are both liberating and oppressive as they represent a narrow set of cultural assumptions about gender. My reliance on technofeminism provides a more global historical trajectory of women's use of and use by technologies of gender, labor, and leisure, both prior to and as part of the current Web 2.0 era, where the binary between the real and the virtual, between rhetoric and reality, is increasingly difficult to discern. Technofeminist research has often combined articulation of women's experiences with an activism that rejects any definition of research that presumes neutrality; instead, such researchers, myself included, aim not only to describe but also to advocate transforming the conditions of those individuals and groups studied. Speaking of such transformation, Faith Wilding (1998) further asserts that "The problem . . . is how to incorporate the lessons of history into an activist feminist politics which is adequate for addressing women's issues in technological culture" (p. 9). For technofeminists, such activism must also acknowledge the role gender, race, class, sexuality, and other identity markers play in the formation of technological ecologies, not as an essentialist, biological variable but as a materialist, cultural, and intersectional one. As a result, technofeminist researchers must interrogate how and why women of diverse backgrounds and identities rely upon and interact with technology in their daily lives, and the cultural contexts that shape and are shaped by that relationship, questions that call for a particular set of methods that enable opportunities for women to share their lived experiences and to have such experiences represented across media genres.

It should be little surprise that a large component of technofeminist work across the disciplines has relied upon narrative, including researcher narrative, as a methodology for generating knowledge about how and why women use and are used by a range of technologies within the larger culture. Wajcman (2001) contends that rather than returning to the either/or rhetorics of technological liberation or technological oppression, "presenting a diversity of narratives . . . enables us to transcend once and for all the traditional dichotomy of technology as either empowering or disempowering for women" (p. 8). For Wajcman, biography as an intersectional methodology foregrounds "the way our experiences are filtered through differences such as race/ethnicity, class, sexuality, and generation" (p. 8). Based on that emphasis on narrative and story, technofeminism is implicitly aligned with third wave feminism and the need to chronicle diverse stories. In her 2015 preface to the updated edition of her canonical *Ain't I a Woman*, bell hooks writes of her early struggle to find a space to be heard: "My experience as a young black female was not acknowledged. My voice and the voices for women like me were not heard . . . I had to explore beyond the classroom, beyond the many treatises and books my fellow white female comrades were creating . . . to offer new and alternative radical ways of thinking about gender and women's place" (p. x). This project attempts to move within and between academic, popular, and industry discourse, aligning theory with lived experiences of women innovators.

In my earlier work (Blair & Tulley, 2007), I have frequently relied on Shumalit Reinharz's (1992) *Feminist Methods in Social Research*. Rather than deploy an objective, prescriptive approach to feminist methodologies, Reinharz articulates a number of flexible tenets that align with *Technofeminist Storiographies*, including that feminist research aims to create social change, represents human diversity, represents a perspective as opposed to a specific method, and perhaps more importantly includes the researcher as a person. Given this latter emphasis on researcher perspective, I strive to make my voice in this project more authentic, sharing both personal and professional experiences involving technology and girlhood culture to establish a sense of positionality. I also strive to maintain what for Reinharz and others is a triangulation of theoretical, historical, and applied methods to address the ways women have been marginalized and excluded from the cultural ecologies both historical and contemporary that comprise computer culture across genres and industries, from films, television, toys, games, and comics, to Silicon Valley itself.

Other technofeminist researchers have adopted Reinharz's framework; for example, Allegra Smith (2016) remixes Reinharz for the digital age, calling for technofeminist methodological tactics that include (re)presentation of marginal lives, perspectives and communities, (re)valuing digital stories and practitioner theory, and (re)ciprocity through participatory design, and reflexivity by mapping positionality, just to name a few. For Smith, the goal of such a remix of feminist methods is to "challenge institutional hierarchies, and to amplify the voices of those typically underrepresented or underserved." Another important aspect of Reinharz that Smith foregrounds is the emphasis on transdisciplinarity, something that resonates with me in my own identification as a feminist and as a digital writing and rhetoric specialist whose focus has been on the politics of technological literacy acquisition for those groups whose access has been mediated by inequitable power-knowledge relations. As a result, writing about the cultural representation of women and information technology is inherently interdisciplinary, connected to conversations in media and communication studies, women's studies, cultural studies, historiography, and others focused on the material conditions that enable and constrain women's personal and professional lives. Given this book's focus on women, information technology, and the cultural representation of the relationship between the two, it is natural to rely upon contemporary current events related not only to gender and technology within Silicon Valley but also the media and popular commentary in trade and other popular publications about those gender, not to mention race and class, dynamics. Indeed, feminist activism is communal and public, relying on the social media, crowdsourcing tools of the Web 2.0 era and making these initiatives visible as they call for accountability and change across industries in ways that align with a fourth wave feminist deployment of social solidarity and powerful coalition building.

Moreover, Ealasaid Munro (2013) contends that there may be a gap between feminist theory and feminist action and that "academic feminism is arguably guilty of failing to properly examine the shape that the fourth wave is currently taking" (online), and while there are numerous interdisciplinary examples of critical analysis of the various forms of digital and "hashtag activism," I concur with the need to bridge the gap between the academy and the public sphere. bell hooks (1981/2015) asserts in her re-assessment of *Ain't I a Woman* of the need to mate her radical politics with her "urge to write . . . and to create books that could be read and understood across different class boundaries" (p. xi). With that goal in mind, the readership for *Technofeminist Storiographies* is naturally broad, academic and independent scholars, popular and professional audiences, as it attempts to provide a series of snapshots in the history of information technology that reflect women's roles and representations (or lack thereof) of those roles; as such these snapshots reflect moments in time that are bound to particular value systems about women's place in society. Rather than suggest that cultural assumptions limiting women's role in technoculture are long behind us, I include numerous contemporary examples, including the larger political context of the United States from the 2016 presidential campaign to hashtag activism movements that include #MeToo. The oppression of diverse women across time and space is ongoing, with the relationship between the larger culture and the culture of information technology being a reciprocal one both then and now. In her 1991 book *Feminism Confronts Technology*, Judy Wajcman documented an early ongoing problem:

> Even the most perceptive and humanistic works on the relationship between technology, culture, and society rarely mention gender. Women's contribution have by and large been left out of technological history. . . . The history of technology represents a prototype inventor as male. So, as in the history of science, an initial task of feminists has been to uncover and recover women hidden from history who have contributed to technological developments. (p. 15)

It is this point that drives *Technofeminist Storiographies*, though admittedly, foregrounding women throughout a Western history of information technology, from Ada Lovelace, to Hedy Lamarr, to Grace Hopper and others, does not automatically change the culture to better accommodate women, as today's tech workers are all too often forced to acclimate to continually sexist, racist, and heteronormative ecologies within the culture of the IT innovation economy. While labor data is an important part of the story of contemporary computer culture, quantitative data alone are an incomplete representation of power dynamics that disenfranchise others. In her *Financial Times* article, "Silicon Valley Upgrades Culture for LGBT Tech Workers," Hannah Kuchler (2017) notes that "many larger tech companies publish data only on sex and ethnicity," even further isolating LGBT employees and obscuring their unique and difficult experiences in adapting to Silicon Valley's techbro culture. Kuchler's reporting highlights the intersectional nature of cultural bias; although Silicon Valley companies have taken very public standards in support of the civil liberties of LGBTQIA and notable coming out stories such as Apple CEO's Tim Cook's (2014), the data that do exist suggest that even for this population, discrimination falls along gender and racial lines, with lesbian women receiving fewer venture capital dollars as opposed to gay males, and LGBT tech entrepreneurs of color faring far worse. Lesbians Who Tech founder Leanne Pittsford has advocated for diverse mentors to foster a sense of inclusive community but also to represent a diversity of perspectives regarding innovation. Focusing specifically on entrepreneurs, Pittsford asserts that "It's really important for an entrepreneur who is starting their own business to have role models and visibility. When you can't relate to someone's lived experience, hiring them feels like more of a risk. It creates friction, but often times friction is what creates the best product" (Chapman, 2018).

Perhaps the greatest challenge in writing a book about information technology is that the culture of contemporary Silicon Valley is one that both shapes and is shaped by assumptions and representations that are simultaneously static and dynamic. This point is equally true of the larger culture as well. I began this book believing that most media and cultural representations of women, girls and technology would be negative; in conducting this applied research, I found that not always to be the case. Despite deep-seated, normative values that continue to erase women and people of color from the rise of computer culture, and despite continued concerns about the limiting depictions of women's lives and abilities on the big screen, I discovered numerous positive portrayals of women and girls, a hopeful trend indeed, but one that doesn't mean that the values of exclusion and marginalization of women and diverse others working in IT are overturned. Some attribute this erasure to a pipeline problem; we don't have enough women seeking degrees, pursuing and maintaining careers, but as I stress, both the rhetoric and the reality of contemporary information technology culture represent an ecology where women struggle to survive longstanding misogyny and sexism, and how the inevitable impact of this oppression on their social and economic well-being keeps them from thriving.

Overall, this book was written over a three-year period from 2015 to 2018, and rather than erase that chronology, I have deliberately chosen to foreground the timeframes of my writing process to better document the relationship between these cultural moments and the various current events relating to women in general and gendered representations of technology in particular. This naturally includes the 2016 United States presidential campaign and the impact of Donald Trump's election on contemporary identity politics. As I write the final sentences of the book, we are seeing a substantial level of feminist and intersectional activism, including increased numbers of women running for political office in response, to counter status quo responses to what are perceived to be the erosion of civil liberties among many diverse, multicultural populations. This includes gubernatorial races; in 2018, Stacey Abrams became the first African-American women voted in as the nominee for governor of Georgia, and Christine Hallquist became the first transgender nominee for governor of Vermont (Greenblatt, 2018). Both are Democrats.

In "The Pink Wave Makes HerStory: Women Candidates in the 2018 midterm elections," the Brookings Institution notes that many journalists have marked these unprecedented numbers as "the pink wave," a greater percentage of women non-incumbent candidates running for office and winning their primaries. Throughout this project, I do critique the cultural alignment of the color pink in product packaging for women and other media artifacts with the presumed performance of femininity at the expense of alternative academic, professional, and social paths for women, including the pursuit of STEM careers. Granted, some uses of the color do have strong feminist goals, as evidenced by the "Pussyhat Project" associated with the 2017 Women's March on Washington, a type of visual political statement. The subtitle of the Pussyhat Project website is "Design Interventions for Social Change," and the organizers note that "Pink is considered a very female color representing caring, compassion, and love—all qualities that have been derided as weak but are actually STRONG. Wearing pink together is a powerful statement that we are unapologetically feminine and we unapologetically stand for women's rights." Even feminist theorist Roxane Gay (2014) acknowledges pink to be her favorite color in *Bad Feminist*, a recognition of her own and what is certainly our own complicity with and resistance to gender performance in our culture, performance that is ingrained virtually at birth, something I discuss in chapter 4. Regardless of the support of the strides made for women in the Brookings Institution post about women political candidates, the reinforcement of gendered semiotic norms, including the color pink, shows that we have a long way to go in counteracting gendered assumptions and making legislative, and certainly technological, workplaces more gender and racially balanced.

In July 2018, Attorney General Jeff Sessions gave a speech to a group to high school students who began to chant, "Lock Her Up," the common mantra aimed at Hillary Clinton during and after her presidential campaign in response to continued allegations of scandal, corruption, and illegal activity past and present. Two years after the election, the chant continues to be common among Clinton detractors, yet as Steve Almond (2018) concludes in "The Disturbing Evolution of 'Lock Her Up'," the rallying cry among Donald Trump's base of supporters is less about any specific crime and more about misogyny against Clinton and others:

> At this point, it's more like a Pavlovian response, generated by citizens whose political views and behaviors are driven not by any coherent ideology, but by a reflexive hostility toward all women. I can hear now much more clearly, in this despotic chant, the desire to create a culture in which men have legal dominion over women and girls. Sometimes this desire is overt. Women and girl migrants who come to America fleeing danger? Lock them up. Women who want to exercise their reproductive rights? Lock them up. Woman who dare to speak about sexual harassment and abuse? Well, if we can't lock them up, we can at least shut them up. (Almond, n.p.)

Given what has been longstanding media inattention to women's achievements, one woman who has clearly bucked that trend is Associate Supreme Court Justice Ruth Bader Ginsburg. Based on her notable dissents, along her status as the second woman after Sandra Day O'Connor appointed to the court, Ginsburg has become a popular culture icon, complete with bobble-head doll. As I complete this project, the documentary *RBG* has been released. Felicity Jones, who earlier portrayed Jane Hawking in the film *The Theory of Everything* that I discuss in chapter 1, will portray Ginsburg in the late 2018 biopic *On the Basis of Sex*, foregrounding the Justice's commitment to gender equality in her life and career. Meanwhile, the visibility of distinguished women scientists and mathematicians does not approximate the celebration of a Ruth Bader Ginsburg. To combat this ongoing lack of recognition, Jess Wade, a British plastics electronics researcher at Imperial College London's Blackett Laboratory has written over 270 Wikipedia pages about women scientists (Devlin, 2018). Wade's efforts are a notable example of technofeminist "backtalk" to those master narratives of male innovation and gender stereotyping, including the European Commission's campaign to promote women and science: "Science: It's a Girl Thing." The campaign includes a video that looks and feels like a music video, complete with hypersexualized young models strutting to the beat and distracting a male scientist from "his" work, and the focus of the women's technical expertise is upon gendered products like lipstick and nail polish.

Whether it's the political arena, the scientific community, or various contexts within today's computer culture, including comics and games, there can be risks involved with both increased visibility and activism. Wade shares a story of being critiqued for bias and favoritism, yet as many of the examples I share throughout the book prove, the backlash can be far worse for attempting to advance professionally and for speaking out about the barriers to that advancement, not to mention that the impact on women's well-being for attempting to assimilate is far more damaging. Silicon Valley has certainly fostered a culture of assimilation and silence in ways that match the quantitative data about the lack of diversity with the qualitative data about the impact of individuals' perennial minority status. In "Going to Work in Mommy's Basement," Sarah Sharma (2018) explains the rise of "broculture," which I discuss in multiple chapters, as the search for a post-mom economy where the adult male children of Silicon Valley have designed apps that replicate a maternal, caregiver mandate that is sexist, racist, and classist and represent their own values and that of the larger culture:

> Peek inside the workplaces of most successful tech companies and you will find ping-pong tables, napping pods, and bottom snacks. That is, these spaces have a striking resemblance to Mommy's basement . . . a culture that "forecloses the possibility of a reconfigured technological future that is not based on the exploiting the labor of others. (Sharma, n.p.)

In this way, today's technologies, as with the household tools Kramarae's collection overviewed three decades earlier, have as much power to disenfranchise those whose roles have been deemed through their labor to be socially and economically subservient. Yet Sharma concludes the role that gender plays in tech is misunderstood to be solely about the absence of women in technology and that "misogyny and racism will not be solved by by diversity in hiring and the inclusion of women alone" but by understanding that the types of technologies that come out of Mommy's basement are designed and coded with an ideology that oppresses rather than liberates women, people of color, and those who identify as LGBTQIA. Sharma suggests it is instead about pervasive power and control that permeates all that we do in that "technologies alter what it means to be human what it means to be in relation to each other." Thus, the culture of the techbro is for the males who thrive within it in a natural one in which gender and racial separatism are the longstanding norm and thus harder to overturn by increasing numbers alone. Regardless of her acknowledgement of the efforts to be more inclusive, Sharma concludes that "The industry is not a monolithic enterprise of techbros—but the future depends on more than representation. . . . Accounting for gender and diversity in the tech industry means contending with the normative regimes of care built into our technologies" and that includes acknowledging the disproportionate systems of power that oppress those who provide that care.

Such concerns connect to technofeminist concerns, for as Wajcman and Le Ahn Pham Lobb (2007) contend, "Technology is . . . understood as a source and consequence of gender relations . . . gender is constitutive for what is recognized as technology, and gendered identities and discourses are produced simultaneously with technologies" (p. 4). Wajcman and Pham Lobb's focus on gender and IT in developing countries, specifically Vietnam, documents the continued devaluation of women's labor, with "low status jobs in a limited number of occupations" and that even when women have the same IT qualifications as men, their employment opportunities are not the same, limited to coding and testing and other support positions in comparison to men who occupy design and specification roles. This suggests a gendered division of labor that is based on differential pay, training, and career advancements, tied less to technical skill rather than to employer perceptions of women's versus men's suitability for specific roles (p. 23), a phenomenon that, as this book highlights, transcends cultural and historical contexts. Wajcman and Pham Lobb call for this perception to be challenged and transformed, and although their article's focus is on global information technology labor data, it is clear through their research that these gender dynamics are replicated in and representative of many but not all developing and developed nations.

On August 9, 2018, Google's daily Doodle featured Mary Golda Ross (1908–2008), the first Native American female engineer who worked for the Lockheed Corporation and later the Skunk Works project affiliated with

NASA. The great granddaughter of Cherokee Chief John Ross (1790–1866), Ross is referred to in a NASA Profile page to be a "hidden figure," in part because her many technical contributions were, like other women, part of classified government activity. In her later years, Ross was a strong advocate for engineering education for women and Native Americans. Yet contemporary statistics document the plight of indigenous women, nearly 85 percent, who have experienced some form of physical and sexual violence. This fact has led to online activist campaigns such as #NotInvisible, organizations that include Canada's National Inquiry into Missing and Murdered Indigenous Women and Girls, and the introduction of U.S. Senate Bill 1942: Savanna's Act, by North Dakota Senator Heidi Heitkamp (Racine, 2017) to reform the law enforcement and legal systems in place for indigenous women such as Savanna LaFountaine Greywind, who was murdered by her neighbor by having her unborn child cut from her womb. In *Hunger: A Memoir of (My) Body*, Roxane Gay (2017) writes powerfully of sexual violence in her early life, situating her story as one among many: "I am weary of all our sad stories—not hearing them, but that we have these stories to tell, that there are so many" (chapter 4, section 71, para. 6). Ross's legacy is important on its own, but I mention it in the larger context of a story of continued oppression of women and girls from diverse cultural groups. We need more than just an annual Google Doodle to celebrate what would have been Ross's 110th birthday; we need to balance such visual rhetoric of celebration with the material realities of women's lives that prevent them from having or pursuing educational or career opportunities. Ultimately, by focusing on a Western history of computer culture my intent in this project is to recover and juxtapose the lived experiences of several prominent women in this chronology with larger representations of women's relationships to both technology and societal norms, a call to turn such *his-story* into not just *her-story but their stories,* not sad, but instead powerful and strong, in a manner similar to earlier calls by Vare and Ptacek (1988) in *Mothers Of Invention From The Bra To The Bomb: Forgotten Women & Their Unforgettable Ideas*.

The remaining chapters address this goal in the following ways: chapter 1, "Parallel Lives and the Recovery of Women in the (His)Story of Computing," chronicles the assumptions about gender and reinforcement of technology in both industry and the larger culture as a male enterprise. Additionally, the chapter aligns these assumptions with media and other depictions of women and technology, foreshadowing more in-depth discussions within later chapters. Equally significant, the chapter also documents, in a case study of two female tech innovators, Ada Lovelace in the nineteenth century and actress and inventor Hedy Lamarr in the twentieth, how an emphasis on recovering the stories of such figures impacts our understanding not only of the role of women in the history of information technology but also of the gender dynamics that impact the false assumption that technological innovation is something of stereotypically male genius. As I shall stress throughout this book, the lone genius is a cultural mythology that past and present media portrayals continue to circulate, and is one that sustains a perception of women's work in IT as subordinate. These rhetorics of technological innovation and achievement as predominantly male impact girls' participation in the STEM areas of study upon which the IT industry relies so heavily and the resulting low numbers of women working in the field as professionals and academics.

Chapter 2, "Distinguishing Rhetoric from Reality in Early Computing Culture," examines the rise of the post-world War II commercial computing industry with the portrayal of women's relationship to technology, relying on the 1957 film *Desk Set*, a Katharine Hepburn and Spencer Tracy vehicle that complicates, as their films often did, the relationship between the sexes but also the relationship between gender and computing. This analysis is grounded in a technofeminist methodology that juxtaposes these media portrayals with historical and biographical accounts of women, including Grace Hopper and the six women programmers of the Electronic Numerical Integrator and Computer, better known by its acronym "ENIAC," and NASA engineer Mary Jackson and her "hidden figures" counterparts featured in Margot Lee Shetterly's 2016 book and later film adaptation. Inevitably, these storiographies represent collaborative approaches to technological innovation that talk back to those rhetorics of individual male genius. Chapter 3, "Bridging the Technological Gender Gap On and Off the Screen," also speaks back to those rhetorics by extending the early discussions in my first chapter about the way in which women's versus men's roles with and relationship to technology are documented on the large and small screen, including recent films such as *Jobs*, *The Social Network*, along with HBO's *Silicon Valley*. These particular dramatizations continue to privilege perceptions of male authority and creativity over more realistic collaborations, as the stories of Ada Lovelace and Hedy Lamarr promote, that involve women as ornamental rather than fundamental. Chapter 4, "Gender Play and the Marketing of Misogyny," examines the gendered assumptions about play within numerous media artifacts that circulate within youth culture, including children's and adolescent literature that both honor historical women figures in science and technology and depict technology use in more gender-fair ways. Finally, chapter 5, "Sustaining a Technofeminist Future for Women and Girls," argues that the continued circulation of cultural myths and stereotypes that negatively impact women and girls' opportunities for inclusion and advancement in STEM fields and the IT labor force call for educational and activist initiatives at the local and national levels. Despite the numbers and diversity of these initiatives, this chapter addresses the continued visibility problem for women in IT. Taken together, these chapters call for a story of the computer revolution that more accurately and fairly reflects both historical and contemporary contributions by women and expedites a more diverse, collaborative future, for which there is no more time to wait.

Chapter One

Parallel Lives and the Recovery of Women in the (His)Story of Computing

Winter 2015. It's Hollywood award season in the United States, and among the consistent nominees for best film are *The Theory of Everything* and *The Imitation Game*. Neither film will win the coveted Academy Award for Best Picture, instead losing to *Birdman*, the redemptive story of a washed-up actor long typecast into a superhero role that by the film's end he embodies both figuratively and literally. The former film is the dramatization of renowned physicist Stephen Hawking's (1942–2018) early life and his diagnosis with motor neuron disease, adapted from his ex-wife Jane Wilde's 2007 memoir *Travelling to Infinity: My Life with Stephen,* one of several books Wilde has written to ensure, as she has noted (Burrell, 2014), her role in Hawking's professional journey and scientific legacy. Regardless of the film's acknowledgment of Wilde's considerable demands of caring for an increasingly disabled adult and three children while trying to write a graduate thesis, the story becomes Hawking's, with Eddie Redmayne taking British and American best actor honors for his portrayal of intellectual triumph over debilitating physical adversity. The latter film represents the struggles, both professional and personal, of the World War II Bletchley Park codebreaker, computer science and artificial intelligence innovator Alan J. Turing (1912–1954). Given Hawking's decades long struggle with amyotrophic lateral sclerosis (ALS), and Turing's conviction, court sanctioned hormonal treatment, purported suicide, and eventual 2013 royal pardon for what were in 1950s Britain illegal acts of homosexuality, their respective professional and personal dramatizations in film and print are undoubtedly worthy of these twenty-first-century accolades.

Because we live in a world of digitally mediated, human-machine cyborg identities that drive our individual habits of mind, of work and of play, it is perhaps fitting we give tribute to and acknowledge the struggles of such scientific and technological pioneers, something Walter Isaacson (2014) notes in his most recent book *The Innovators: How a Group of Hackers, Geniuses, and Geeks Created the Digital Revolution.* Although Isaacson includes a discussion of nineteenth-century computer visionary Ada Lovelace, he has admitted that until very recently, he did not know who she was, learning of her contributions from a paper written by his teenage daughter, as Audrey Watters (2015) confirms in her article "Men (Still) Explain Technology to Me: Gender and Education Technology." For Watters, "even a book that purports to reintroduce the contributions of those forgotten 'innovators,' that says it wants to complicate the story of a few male inventors of technology by looking at collaborators and groups, still in the end tells a story that ignores if not undermines women. Men explain the history of computing, if you will."

Undoubtedly, throughout the history of information technology, the stories of revolution that typically make it to the printed page or the big screen are inherently male. Women, from Hawking's wife Jane Wilde to Bletchley Park's Joan Clarke (1917–1996), Alan Turing's intellectually equal colleague but professional subordinate, are portrayed as helpers, historical handmaidens in service of a male-dominated scientific frontier. And both Clarke's niece Inagh Payne and Turing's biographer Andrew Hodges have publicly critiqued the dramatization for overly romanticizing the relationship and for selecting actress Keira Knightley to portray the "rather plain" Clarke (Lazarus, 2013). As with Turing himself, Clarke's full range of contributions is less known because of the secrecy surrounding the wartime activities at Bletchley Park. Even the American film posters for *The Imitation Game*, while including Keira Knightly, positions her as standing by, à la Tammy Wynette's classic country standard, but clearly behind her man.

It would be comforting to believe that in the twenty-first century, these portrayals represent a particular historical moment in Western culture when women's roles were inevitably subservient to men across professions, an era that is presumably long behind us. However, in October 2014, Microsoft CEO Satya Nadella drew controversy for comments about the gender gap in pay, discouraging working women from self-advocacy and indicating that good "karma" would result for avoiding what Facebook CFO Sheryl Sandberg and others contend is an unfair stereotype of the aggressive female employee, a binary opposite to the acquiescent handmaiden, and a workplace parallel to a larger longstanding cultural binary that positions women as either madonnas or whores. The gender gap in pay remained a national priority for then U.S. President Barack Obama; in his 2015 State of the Union Address, Obama called for equity and chiding legislators with the comment that "It's 2015. It's time." It's also time for women's stories of success and entrepreneurship to be shared and heard, for while a January

2015 *Newsweek* report, "What Silicon Valley Thinks of Women," documents women's successes and challenges in securing investors for Silicon Valley startups, the cover image features a graphic of a faceless woman, clad in a short red dress and red stilettos with a computer arrow icon attempting to lift her skirt.

As Nina Burleigh contends in this cover story, "The only problem with their dream is that Silicon Valley has never produced a female Gates or Zuckerberg. There are a few high-profile female entrepreneurs in the Bay Area, but despite the very visible success of corporate titans Meg Whitman of eBay and Hewlett Packard, Sheryl Sandberg of Facebook, and Marissa Mayer of Yahoo, who signed up with companies after they took off—their numbers are relatively minuscule" (online). Yet if Hillary Clinton's 2015 speech and question and answer session for nearly 5000 women at the Watermark Silicon Valley Conference for Women (Chozik, 2015) is any indication, increasing numbers of women in technology are looking for more political awareness of their existence and their plight, as Clinton appeared to demonstrate, in the quest for equal pay, paid leave, and the need to break through the glass ceiling that she herself experienced as a presidential candidate in 2008 and again in 2016. Similarly, Janet Abbate's (2012) *Recoding Gender: Women's Changing Participation in Computing* affirms that STEM fields remain male dominated, lacking sustained professional opportunities for women and girls to enhance both technological aptitude and attitude, despite the historical evidence that women mathematicians played a formative role in the development of the information technology industry and the rise of computer science as an academic discipline.

All too often, when women challenge the cultural conditions that reinscribe traditional gender roles in information technology careers and elsewhere, there is severe backlash. For instance, feminists have consistently critiqued the gaming industry for misogynist representations of and resulting violence against women with the advent of #gamergate, or what Lauren Williams (2015) from ThinkProgress.org describes as a "small subset of the gaming community that have harassed female media critics, developers and bloggers with violent and graphic death and rape threats." Supporters often characterize it as a movement for improving ethics in gaming journalism after game developer Zoe Quinn's former boyfriend publicly accused her of cheating on him with a gaming writer who supposedly favorably reviewed her interactive game Depression Quest.

Sadly, since these early developments in 2014, #gamergate tactics of violent sexist and racist online harassment have extended beyond the gaming community to other aspects of digital and political culture. A key instigator has been Milos Yiannopoulos, the controversial conservative commentator and #gamergate apologist, who was ultimately permanently banned from Twitter for his racist harassment of actress Leslie Jones in 2016 for her role in the all-women remix ensemble of the film *Ghostbusters*. Yiannopoulos has continued to play a role in alt-right political movements even after being removed at the conservative forum Breitbart following the coverage of statements in defense of "consensual" sexual relationships between young boys and adults. Yianoppoulos (2016) has himself described tech broculture as a "sexodus" from co-equal professional and personal relationships with women: "The rise of feminism has fatally coincided with the rise of video games, internet porn, and sometime in the future, sex robots. With all these options available, and the growing perils of real world relationships, men are simply walking away" (Breitbart). Throughout this project, I include multiple examples of the impact of techbro ecologies on women's professional advancement and the personal safety, and in my final chapter, I will discuss Zoe Quinn's recent efforts to create spaces in which current and future targets can find social, personal, and legal support.

The so-called "ethical" behavior of #gamergate has resulted in much more than virtual harassment; these threats, which have involved distribution of home addresses and telephone numbers, have resulted in a real-time disciplining of women's professional and personal words and actions, with renowned activists such as Anita Sarkeesian ultimately declining speaking a speaking engagement at Utah State University in October 2014 after threats of violence, and when the University could not guarantee her and her audience's protection in light of Utah's Open Carry Law. Despite the shifting statistics that indicate that women's participation in gaming environments almost equals their male counterparts by a percentage of 44 percent female and 56 percent male in 2015 (Statistica, 2015), women's lived experiences and narratives of those experiences in these and other technological spaces are diminished, silenced, and left unheard and invisible. Often when they try to articulate their perspectives in social media, as did SendGrid developer evangelist Adria Richards after hearing sexist language and behavior at a largely male-attended tech conference, they experience brutal violent backlash, as Richards and the company she worked for did after one of the men she reported in social media was fired, with Richards suffering the same fate soon after.

With these experiences in mind, this opening chapter acknowledges the challenges women working in the IT industry face, challenges based primarily on assumptions about gender and the reinforcement of technology as solely individual and male rather than as collaborative and female. To highlight the importance of such collaborations, I include the parallel stories of Ada Lovelace and Charles Babbage in the nineteenth century, and Hedy Lamarr and George Antheil in the twentieth to disrupt the historical and contemporary mythos that shapes and is shaped by cultural representations that impact and often subordinate women's lived histories of computer culture.

My argument throughout this book is that despite the rich, complex stories and histories women have in technology—as programmers, inventors, and workers, cultural artifacts such as film, television, games, toys, children's books and biographies, inadequately and inaccurately represent those stories and histories, reinscribing a techno power-knowledge dynamic that has continued to limited women and girls' education and ultimate participation in technological arenas. Inevitably, the material conditions that surround technology use and deployment in local and global cultures impact the extent to which women and girls gain and sustain access within those cultural contexts. In this way, technological empowerment is not only an intersectional feminist issue but also an economic one as well, requiring that women be supported within those STEM fields governing the technological labor force and thus able to contribute to present and future innovation. This support can in turn sustain a documented history of diverse women's technological achievements in which their stories are heard for generations to come, rather than be forgotten and unknown.

INTERROGATING CULTURAL ASSUMPTIONS ABOUT WOMEN AND TECHNOLOGY

As someone who identifies as a technofeminist, it is vital for me to interrogate the consequences of not portraying women's stories, particularly in fostering their continued participation in an ever-increasing technological labor force. The short-term impact may rest in my most recent trip to my local bookstore, and its inadvertent attempts to define but inevitably delimit girl's emerging creativity in alarming ways (figure 1.1).

In this display, "Girls Creativity" hails a number of cultural and semiotic codes: primarily pink, hearts, lots of princesses, along with opportunities for jewelry making. Certainly, there exist attempts by the toy industry to orient girls to technology, including children's books such as the 2010 *Barbie: I Can Be a Computer Engineer*, which portrays the ageless blond as possessing the creative initiative to design a game to teach children how computers function, but ultimately in need of male assistance to both implement the design and remove a virus from her computer. On Amazon.com the book was originally bundled with *Barbie: I Can Be an Actress*. One reviewer on Amazon.com stated, "As a computer engineer and the father of two daughters who are both in STEM fields, my only recommendation for this book would be to set it on fire." Such public outrage against this longstanding narrative of male dominance led not only to backlash from women inside and outside the IT industry but also to a number of counter-narratives from sites like Feminist Hacker Barbie (NPR), and the book was ultimately discontinued by Mattel/Random House. To its credit, Mattel has since developed a

Figure 1.1.   Bookstore Display. Photograph by the author.

Barbie Robotics Engineer Doll available in four different models representing differing ethnicities in a partnership with Tynker to introduce girls to coding and other careers in which programming presumably plays a role, including an astronaut, a beekeeper, a farmer, a musician, and a pastry chef. Mattel's website for the series states that "The STEM field continues to grow, but only 24% of those jobs are held by women. That's why we're encouraging girls to explore STEM through imaginative play and our partnership with Tynker, the award-winning computing platform."

That the original Barbie engineer book was written by a woman is perhaps the greatest indicator that our assumptions about women's proficiency with technology have become a mythology of the larger culture that "goes without saying." This rhetoric is far stronger than any reality of today's women working in technology and the larger contemporary history of women who shaped it, such as Rear Admiral Grace Murray Hopper. Hopper (1906–1992), one of the first computer programmers on Harvard's Mark I, received the first Computer Science "Man" of the Year Award in 1969 among her many awards and countless honorary degrees. Despite accolades that include the Grace Hopper Celebration of Computing and even the christening of the USS Hopper in her honor, in a recent documentary on Hopper,

The Queen of Code, women information technology professionals lamented that "All we talk about is Steve Jobs and Bill Gates," leaving women and girls fewer roles models and "little historical knowledge of women's contributions in the early days." As Megan Smith, former Chief Technology Officer of the United States contends in the documentary, "Grace Hopper is like an Edison but she's absent from the history books." Yet A 2015 National Public Radio story on Grace Hopper's Career notes biographer Kathleen Broome Williams contention that Hopper would have had hated the reference to herself as "The Queen of Code," which given the larger cultural assumptions that place even successful women into gendered roles of princesses and queens, seems inappropriate for a woman who instead referred to herself as a "pirate," and a person who didn't have to think much about feminism because she was "in the Navy." I focus on Hopper's distinguished career in chapter 2.

Although Hopper may have resisted her iconic status as a first among women in the history and advent of computing as a profession and a discipline, it is clear that she was concerned about the need for educating future generations, regardless of gender: "You've gotta get out there and help us train the youngsters. Teach them to go ahead and do it. Teach them to have courage. Teach them to use their intuition, to stick their necks out. You've got to move to the future. We're going to need *all* [my emphasis] of them" (NPR). Admittedly, Hopper did not identify publicly as a feminist. However, the data about the dearth of women in STEM both in the academy and in the professions are strong rationale for a technofeminist recovery of women's contributions, as well as a technofeminist analysis of how those contributions are disseminated in the larger culture.

Hopper's technological history, along with those of her wartime women trailblazers on ENIAC in the United States and COLOSSUS in Great Britain, her predecessor Ada Lovelace (1815–1852), considered to be the first computer programmer, and her mid-twentieth century contemporary Hedy Lamarr (1914–2000), co-developer of the frequency hopping technology during World War II, are all largely absent from the big and small screen. Granted, there are nods to women's roles in technological innovation in the form of the recent *Bletchley Circle*, the dramatization of four female codebreakers who come together in post-war London to solve murders, disguising their collaborative sleuthing as a women's bookclub. This is hardly the stuff of technological legend, as we see in the form of films like *Jobs*, where the legendary title character is portrayed by the bankable box-office heartthrob Ashton Kutcher. In a 2015 interview on *Charlie Rose* (Siede, 2015), Megan Smith contends that even though women were part of the original Apple Macintosh team, none of them were represented with speaking roles in the film. This phenomenon is not only due to gender but also due to a cultural privileging of conceptions of individual genius and innovation over the more collaborative models that I contend a technofeminist historiography can inevitably uncover to document women's important roles. The limited presence of women innovators on screen, however, is part of a larger problem identified by San Diego State University's Center for the Study of Women in Television and Film, where they report that "only 12 per cent of protagonists in the top 100 highest-grossing domestic films of 2014 were female." The Center concludes that "in assessing 2300 characters in 100 films ...women were more often cast in supportive roles where they helped others" (McDonald, 2015), something viewers see in films that include *The Theory of Everything* and *The Imitation Game*. This technological absence extends to television as well, most recently in the form of the HBO series *Silicon Valley* that I discuss more extensively in chapter 3. The series features six male tech employees as they begin a start-up company, and which has been critiqued for male stereotypes that seem to compete for ratings with the tech-nerdiness of the characters on programs like *The Big Bang Theory*.

Such concerns also impact female actors as well. In her 2015 Golden Globe award acceptance speech for her role in the television miniseries *The Honourable Woman*, Maggie Gyllenhaal (Dreyfus, 2015) made a strong case for documenting the lived experience of women on screen:

> I've noticed a lot of people talking about the wealth of roles for powerful women in television lately. And when I look around the room at the women who are here and I think about the performances that I've watched this year what I see actually are women who are sometimes powerful and sometimes not, sometimes sexy, sometimes not, sometimes honorable, sometimes not, and what I think is new is the wealth of roles for actual women in television and in film. That's what I think is revolutionary and evolutionary. (Dreyfus, n.p.)

Certainly, as Gyllenhaal suggests, there needs to be a focus on more "actual" women in film; more positive examples include the award-nominated performances of Reese Witherspoon as Cheryl Strayed in *Wild*, and Laura Dern as her cancer-stricken mother, along with the award-winning performances of Julianne Moore as a successful college professor diagnosed with Alzheimer's disease in *Still Alice* and Patricia Arquette as the frequently single mother of two children in the twelve years of making the film *Boyhood*. Yet even these example prove problematic in their lack of female diversity.

Even with these partial triumphs, however, female leaders and innovators have been depicted as subordinate in history, on screen, and through various texts and artifacts that circulate that patriarchal ideology. Innovators such as Grace Hopper have been recognized and lauded, but nevertheless, her story fits into a history of information technology that is not only gendered but also raced and classed, especially when considering the contributions of women of color such as Shirley Ann Jackson, the current president at the Rensselaer

Polytechnic Institute. Jackson is the first African-American woman to earn a doctorate from the Massachusetts Institute of Technology, and one of the few female recipients of the National Science Board's Vannevar Bush Award for public contributions to science and technology. As a physicist, Jackson's impact on telecommunications research has enabled innovations that include the portable fax, touchtone telephone, and fiber optic cables. Clearly, Jackson's contributions to the IT history are well documented when compared to the efforts of Marie Van Britton Brown (1922–1999), a New York City nurse credited, along with her husband, of patenting and inventing the closed circuit television systems that are so much a part of our security and surveillance culture. In chapter 2, I shall also discuss the visibility, through Margot Lee Shetterly's *Hidden Figures*, of the three African-American women who played such a vital role in the U.S. space race at NASA. Despite these contributions, the numbers of women of color employed in the IT industry remain low. As Gail Sullivan reported in a 2014 *Washington Post* article on Google's diversity data in relation to larger Bureau of Labor Statistics percentages, only "four percent of employed software developers in the United States are African American, 5 percent are Hispanic and 29 percent are Asian. . . . Comparatively, 1 percent of the Google's tech workforce is black, 2 percent is Hispanic and 34 percent is Asian" (online). Although Google's data are parallel to other Silicon Valley giants, it, along with Facebook and Yahoo, has been more forthcoming about its lack of improvement in the area of diversity (Hu, 2014).

Given the limited number of women, particularly women of African-American or Hispanic-American descent, working in the United States IT industry in technical or designer roles, it is not surprising they would potentially feel isolated and subject to less than hospitable work environments. Such isolation and lack of advancement opportunity for women is a factor Facebook CEO Sheryl Sandberg has written about extensively, though from a privileged race, class, and, according to Roxane Gay (2014) heteronormative position, in her 2013 book *Lean In: Women, Work, and the Will to Lead*. Sandberg acknowledges the double-bind facing women, needing to be assertive but ultimately stigmatized as difficult and less of a "team player" for doing so. Sandberg shares a range of stories to show the challenges and successes for women seeking leadership roles in the IT industry and beyond. And while I will contextualize her book within the context of Silicon Valley culture, both real and dramatized, in chapter 3, as well as offer my own reaction to it as a reader, it is important to remember its limits from an intersectional perspective. Roxane Gay concludes that Sandberg's book "cannot and should not be read as a definitive text or a book offering universally applicable advice for all women, everywhere. . . . Perhaps we can consider *Lean In* for what it is—just one more reminder that the rules are always different for girls, no matter who they are and no matter what they do" (p. 313).

Of equal concern are reports of sexism, misogyny, and harassment, factors that in 2014 led nine women in the tech industry to post a response at "About Feminism." The site, referred to as a manifesto by blogger-journalist Cory Doctorow (2014), represents a powerful statement about why feminism, far from being a dirty word in the larger culture, is a necessary form of political action for women working in IT to counter what they see as the lackluster efforts by Google and others to more diverse hiring and retention practices:

> We have watched companies say that diversity is of highest importance and have invited us to advise them. After we donate much of our time they change nothing, do nothing, and now wear speaking to us as a badge of honor. Stating, "We tried!" We've grown cynical of companies creating corporate programs and paying lip service to focusing on women's issues in the tech industry without understanding the underlying reality.

Admittedly, the word "manifesto" is appropriate, for in the Marxist sense, the nine authors of "About Feminism" want to do more than critique. Instead, they want to encourage transformation, particularly through involvement and volunteerism in organizations that educate and prepare women and girls, especially minorities, for careers in the tech industry and to understand that the emphasis on women in IT is not just an emphasis on gender:

> While this letter speaks specifically about our experiences as women in tech, to build true diversity in tech we must address more than one aspect of gender, more than any one aspect of our identities. Our efforts must address, and be inclusive of, race, class, sexuality, gender identity and expression, and their intersections.

Other responses are equally if not more radically activist in their call to action. In her *Guardian* article "Screw Leaning In. It's Time to Slam the Door in Silicon Valley's Face," Jess Zimmerman (2015) reports on the site "tableflip dot club" and its call to action for women working in IT. Several of its key tenets include:

> Women are leaving your tech company because you don't deserve to keep us around.
> Fuck that, we're done. It's not us, it's you.
> When we try to take a seat at the table like Sheryl said we should, we're called presumptuous.
> It's time we take our potential elsewhere. (Table Flip Club, n.p.)

With its own Twitter account, the group advocates not leaning in, but "pulling together," and as Zimmerman's article highlights:

> the difficulties women face aren't the problems of one woman, or one team or one company. "They're not just limited to Adria Richards or Brianna Wu or Ellen Pao; they're not just Twitter with its no-women board or Wikipedia with its 91% male editors. The problems are systemic—and no amount of attitude adjustment or leaning in on the part of those who get screwed by the system can possibly change it. (Zimmerman, 2015, n.p.)

The Barbie Engineering fiasco suggests there is a trickle-down effect between the realities of women in the workforce in general and the technological workforce in particular and the way in which that labor is portrayed in cultural artifacts like toys and games, marketed to children. Nevertheless, there are attempts to foster more "creative" alternatives for girls, most recently in the form of initiatives like GoldieBlox, engineering erector sets and other educational toys, specifically created and marketed to girls by a married couple, Debbie Sterling (founder and CEO) and Beau Lewis, and supported by organizations that include Techbridge, Girls Who Code, Black Girls Code, and the Society of Women Engineers. In a GoldieBlox video marketing campaign that has gone viral, one video shows three girls in front of a television, one African-American, one Asian-American, and one Caucasian bored at the dolls and princesses they are encouraged to play with and emulate in dress and mannerisms. As the commercial voiceover chorus of young girls assert, "you like to buy us pink toys, and everything else is for boys." The GoldieBlox motto is "disrupting the pink aisle." Yet, not unlike big screen portrayals like Keira Knightley's Joan Clarke first as a girlfriend and then as a mother-figure to Benedict Cumberbatch's chemically castrated Alan Turing, even initiatives like GoldieBlox can reinforce cultural assumptions about race, gender, and technology, as the GoldieBlox mascot is a Caucasian girl with long blond hair, and a number of their toys are as "pinkwashed" as Mattel and Lego toys aimed at girls, including Barbie Computer Engineer (Miller, 2013). This suggests that the representation of the STEM workforce is as white as it is male, and ultimately not only gendered, but raced, despite the diverse representation in the video campaign. The GoldieBlox mascot certainly counters the "dumb blonde" stereotype that continues to circulate, and communicates to young girls that computers do not signify the social awkwardness associated with the male stereotype of the computer geek. She also counters female stereotypes such as the brilliant, neuroscientist Amy Farrah Fowler (portrayed by actual neuroscientist and former child actress Mayim Bialik) as compared to the slow-witted but sexually savvy street-smart waitress Penny (who as an "every-blond" has no last name) in CBS's *The Big Bang Theory*. Nevertheless, GoldieBlox also privileges traditional standards of beauty and femininity for a young audience even as it subverts the assumption that STEM activities are for geeky, inept boys and men, as they are equally stereotyped in *The Big Bang Theory* in the form of Leonard, Sheldon, Howard, and Raj. In this way, the GoldieBlox example is simultaneously complicit with and resistant to a portrayal of women's and girls' experiences with and contributions to technological innovation, a process that diminishes their role—past, present, and potentially future—in the larger cultural history of information technology.

Thus, one goal of *Technofeminist Storiographies* is to analyze how the mythos surrounding technology and technology innovators is marketed and packaged for youth consumption and how that is an inherently gendered process across various genres that I include in this project. That such consumption is important is clear from work done by the American Association of American Women, whose reports *Tech Savvy: Educating Girls in the New Computer Age* (2000) and *Why So Few: Women in Science, Technology, Engineering, and Mathematics* (Hill & Corbett, 2010) document how the lack of women role models in schools and universities, along with the cultural bias that mathematics is a male domain, limits girls' pursuit of advanced mathematical training in school, are not so subtle indicators that STEM areas are for boys only. With the limited change in statistics for women working in STEM and the continued male profile of the computer-scientist working in the IT industry, it is vital to look at the cultural circulation of technology narratives that empower men and boys and disenfranchise women and girls, including images that objectify women in the promotion of technological access and literacy. This includes sites such as codebabes.com, which advertises, presumably to a male clientele, the ability to learn code and check out babes simultaneously. Notably, a counter-site titled codedicks.com represents a response from a male segment of the tech industry and disclaims that "CodeDicks is in no way affiliated, nor condones the beliefs of CodeBabes. This is satire. None of the dudes on this site approve of sexism and would like to see people of all genders treated equally and respectfully, particularly in our industry."

Speaking of his motives behind his 1972 book *Mythologies*, Roland Barthes famously states in his preface:

> The starting point . . . was usually a feeling of impatience at the sight of the "naturalness" with which newspapers, art and common sense constantly dress up a reality which, even though it is the one we live in, is undoubtedly determined by history. In short, in the account given of our contemporary circumstances, I resented seeing. . . . History confused at every turn, and I wanted to track down, in the decorative display of *what-goes-without-saying*, the ideological abuse which, in my view, is hidden there. (Barthes & Lavers, p. 11)

Similarly, my own motives are to not just uncover, but recover women's roles in the history of technology and to redress "what goes without saying," in ways that make visible the cultural and rhetorical processes of disseminating technological narratives through film, television, advertising, visual images, Web 2.0, and even displays in the local bookstore. I also wish to demonstrate the way our cultural assumptions about gender and technology continue to leave insufficient space for women and girls to become part of a larger cultural and economic present and future. This is not an attempt to diminish the contributions of men, past and present, in the dynamic history of information technology. Rather, it is an attempt to make visible the mythos surrounding technology and in doing so help men and women confront and potentially transform them. While such efforts are evident in the national Computer History Museum's efforts to include women leaders and innovators in their exhibits, the larger cultural history of women and technology positions those contributions as peripheral.

Although many media portrayals of women suggest subordinate roles—as the most recent *Imitation Game* film clearly suggests—even the programming of technology itself is a gendered process. In "Rise of the Fembots: Why Artificial Intelligence Is Often Female," *LiveScience* staffwriter Tanya Lewis (2015) reports that many Artificial Intelligence (AI) systems have female personae, in part as one interviewee indicates that the roles such systems fulfill are subordinate, e.g., maids in high-tech hotels, guides in museums. Certainly, this is not new; although the male Robby the Robot made his film debut in the 1956 film *Forbidden Planet* and went on to appear in a wide range of science fiction television programs, other examples include Rosie, the space-age housekeeper on the popular 1960s primetime cartoon the *Jetsons*, and the more sinister android replacements in the 1975 and later 2004 film *The Stepford Wives*.

Donna Haraway's (*Simian's*, 1991) "Cyborg Manifesto" may have envisioned the cyborg as the hybrid between human and machine and a preferable state for women than that of the "goddess." However, artificial intelligence, when programmed by men to be female, assumes subservient and thus inferior roles. As Lewis reports, Apple's original Siri, whose Nordic name literally means "a beautiful woman who leads you to victory," was only a female voice, though it is now possible to select a masculine voice as well. As one AI expert contends in Lewis's report, robots and other forms of AI may be programmed as female, even childlike, so that humans become more comfortable around them and less threatened by them, as we see in such classic portrayals such as Hal, the docile computer gone rogue in *2001: A Space Odyssey*. Despite his status as a cultural icon, in today's big screen, high tech, sci-fi productions, Hal is the equivalent of the early Atari video game Pong, representing a rudimentary artificial intelligence, particularly when compared to films such as *Lucy*, starring Scarlett Johannson as the namesake of the world's first woman in prehistoric times. Johansson is not new to AI performances, given her role as the operating system who loves and then leaves Joaquin Phoenix's lonely, introverted character in the 2013 film HER. In the case of Lucy, however, Johannson's character moves from ditzy to all-powerful as she is surgically implanted with a nootropic chemical that, when accidentally distributed in her system, enhances her tele- and psychokinetic powers with each scene to morph from human to machine, a large drive that contains the knowledge of the universe with her final words via text message "I am everywhere." Inevitably, Lucy transcends the restrictions of gender that made her the prey of male druglords and the research subject for male scientists.

Regardless of such gender transcendence via AI, contemporary advertising wars, including between Microsoft and Apple, play up on gender distinctions. For example, the 2013 campaign for the Windows Surface RT positions the iPad as inferior because of its lack of various features and its higher cost. The iPad's Siri asks the concluding tag question "Do you still think I'm pretty?" in a stereotypically gendered fashion. Inevitably, there are historical connections as well that align the use of certain technology with women, and in doing so demean them or engage in a process of de-skilling the labor and knowledge associated with them, such as when telephone operating became women's work, or women's tools for gossip . . . or too much talking, which is implied with the end line of the Surface commercial. The ad is inevitably a contemporary artifact that promotes this cultural myth of gender and communication, with earlier games and toys for girls such as "Girl Talk" and "Girl Talk Dream Phone." I discuss such childhood artifacts in greater detail in chapter 4. Within this cultural context, the iPad becomes "girly" when compared to the efficiency of the Surface RT.

Earlier campaigns between Apple and Microsoft also position women problematically, even peripherally to the personification of technology as male, as the Apple "Get a Mac" campaign of "I'm a Mac vs. I'm a PC." The campaign, which ran from 2006–2009, features the well-known male characters of the cool, casual MAC dude in jeans and a T-shirt and the stuffy, awkward, and less attractive male PC in his business suit, who argues his relevance and inevitably loses to his hipper counterpart. When women do appear in the campaign, they typically are represented as peripheral, such as a Japanese woman with a digital camera. In one version, when MAC and PC are comparing the desktop video editing tools I-Movie and Moviemaker, MAC's movie appears as the supermodel Gisele Bündchen, while PC's movie appears as a man dressed and bewigged as Bündchen's clearly less attractive doppelganger. In this campaign, PCs aren't hip, cool, or attractive, and the presence of Bündchen and her doppelganger are not as users of technologies but objects of it. Such messages about technology and women's relationship to it, as they circulate in film, television, advertising, and toys, are complicated ones. Undoubtedly, these artifacts simultaneously distribute and create for current and future generations of women a narrow set of gender roles and expectations.

RECOVERING ADA LOVELACE AND HEDY LAMARR

Just as girls today continue to experience gender norms that privilege physical beauty and the duties of wife and mother that subordinate independence and intellectual achievement, so too did two of the most notable—but until the later twentieth century largely ignored—female innovators of the nineteenth and mid-twentieth centuries, Ada Lovelace and Hedy Lamarr. Born ninety-nine years apart, in 1815 and 1914 respectively, and into very different societal contexts, Ada Lovelace and Hedy Lamarr's parallel lives are not merely tied to their respective contributions to information technology. As the daughter born to the romantic poet Lord Byron and his wife Anna Isabella Byron, and with months to be a child of divorce between the two, Augusta Ada Byron's eventual contributions to the earliest histories of computer science, given her designation as the first computer programmer, are certainly feted within information technology circles. This includes a programming language named after her in 1974, though such recognition occurred more than a century after her untimely death from uterine cancer in 1852. And the Austrian born Hedy Eva Maria Keisler, who would herself evolve into the "most beautiful woman in the world" on the silver screen and later in life received recognition from organizations such as the Electronic Frontier Foundation for her collaborative work on frequency hopping, or spread spectrum technology, the precursor to today's wireless and cell communications. Unlike Lovelace, Lamarr lived a long life, dying in January 2000 at the age of 85, having made it into a new century largely forgotten as an actress, but with contributions to information technology that until very recently were eclipsed in the larger culture by her career as a femme fatale on and off screen. Biographies of both Lovelace and Lamarr abound, including recent treatments such as James Essinger's (2014) *Ada's Algorithm: How Lord Byron's Daughter Ada Lovelace Launched the Digital Age* and Richard Rhodes's (2011) *Hedy's Folly: The Life and Breakthrough Inventions of Hedy Lamarr, the Most Beautiful Woman in the World.*

Perhaps what unites Lovelace and Lamarr as technological pioneers is in how, as Walter Isaacson (2014) concludes "the truest creativity of the digital age came from those who were able to connect the arts and sciences" (online). Further investigation into the lived experiences of these two women documents the tension between the arts and sciences in both personal and political ways. For Lovelace, mathematics represented the polar opposite from the romantic escapades, both real and imagined, of her infamous father, Lord Byron, who lived the type of passion-filled life he committed to print.

Isaacson notes Byron's early humor toward and later derision of wife Annabella's status as "the Princess of Parallelograms," and as a "walking calculation." It was that derision, and the fear that daughter would be just like father, that shaped Annabella's decisions to promote science and mathematics as a primary focus of young Ada's study, sponsoring a passion for computational literacy that stuck through Lovelace's brief but eventful life. For many young women of Lovelace's era, marriage was the top priority for socioeconomic advancement in a culture in which titles were as important as the creature comforts that accompanied them.

Both Ada and Hedy made good marriages (though Hedy made six trips to the altar), yet both would experience the restlessness and sense of entrapment that the expectations of marriage and second-class status engender. Ada's 1835 marriage to William, Lord King was one that pleased her mother, who insisted that her daughter form an aristocratic union that garnered a title. Once he elevated in peerage to the Earl of Lovelace, and once Ada had delivered three children within four years of her marriage, she resumed focus on her mathematical studies, working with renowned mathematician Augustus de Morgan (1806–71). All the while she battled an un-diagnosable illness that confirmed the misogynist assumption by de Morgan and the larger society that women in general, and in this case Ada in particular, were too physically and intellectually delicate for such intense scientific study (Dodson Wade, 1994). As a result, scientific doors were typically closed to women, despite the fact that Ada's analytical thinking frequently went beyond the knowledge of childhood tutors and mentors. And Hedy's story is no less compelling than Ada's; as an only child, she grew up adored by both parents, particularly her father who encouraged her interests in science. Thus they were alarmed at Hedy's foray into acting, fostered in part by her adolescent transformation into a dark-haired, alabaster-skinned beauty, that had her abandon her studies early on and make one of her first screen appearances in *Ecstasy*, an ironic art-depicting-life story of a restless young wife who seeks passion outside marriage and a role that called for Hedy to appear nude.

The history of women's role in information technology is, not surprisingly, complicated, in part because of the cultural expectations and overall representations that obscured and devalued the labor of these digital pioneers. Moreover, this early and continuing devaluation has been constant for nearly two centuries: produced, distributed, and consumed in the larger culture through genres including advertising to film and newer media. I include the parallel lives of Ada Lovelace and Hedy Lamarr as women whose intellects and resulting innovations and inventions were inevitably thwarted by their historical moments, including material lack of tools that could have enabled the inventions they envisioned. Yet what also makes Ada's and Hedy's stories compelling and again complicated from a technofeminist standpoint is their intellectual partnerships with men, Charles Babbage (1791–1871) and

George Antheil (1900–1959), that document co-equal rather than subordinate relationships such as those portrayed in dramatizations like *The Imitation Game*. In fairness, the failure of the industrial revolution of nineteenth century and the digital revolution of the twentieth to catch up to the creative genius of Lovelace and Lamarr impacted the work of Babbage and Antheil. Perhaps the vulnerabilities of these particular men and their own thwarted professional journeys also enabled a more reciprocal partnership that for most male-female relationships of their respective world orders were unrealized, from the bedroom to the boardroom.

What brought both pairs together, Ada and "Babbage" (as she came to affectionately call him), and Hedy and George, were similar passions and motivations. If Ada Lovelace's historical legacy is that of the first computer programmer, it is Charles Babbage's status as the developer of the first computer that makes their contributions reciprocal and symbiotic. Born in 1791, Babbage was a mathematician renowned for his design of the "Difference Engine," and first published in a paper for the Royal Astronomical Society in 1822. The invention was able to calculate and print astronomical and mathematical tables, based in part on prior work by the German engineer J.H. Müller (1746–1830) and French engineer Gaspard de Prony (1755–1839). While the British government ultimately funded Babbage's design for over 17,000 pounds, it remained a design, never realized until well into the twentieth century, when it was replicated by and displayed at the London Science Museum. Babbage had already moved on from the Difference Engine's ability to merely calculate to the newer Analytical Engine's ability to store and process punch-card based programs, in other words the precursor to the earliest twentieth-century computers. As James Essinger (2014) suggests in *Ada's Algorithm*, Babbage's status as a member of the Royal Society and a fixture in the early nineteenth-century society circuit drew a large audience for his ideas, including Annabella Byron and her impressionable daughter Ada. But Babbage, like many of his social class, was in no need to make his innovations commercial. Hence ideas like the Difference Engine and Analytical Engine remained that, ideas on paper that the government grew weary of supporting when many years passed without a deliverable, leading to British Prime Minister Robert Peel's refusal in 1842 to fund a machine without a pragmatic value in larger society, beyond the speed and accuracy of calculation, that was unknown, even to Babbage himself.

Not unlike Ada, Babbage's larger contribution to information technology would go unrecognized for close to a century. But where Peel failed to recognize and support Babbage, Ada became his ardent, but when necessary, constructively critical admirer. Recent biographers have attributed Lovelace's passion for the Analytical Engine to her roots as the daughter of the infamously romantic Lord Byron, aligning head and heart in her enthusiasm for Babbage's invention and her ability to see beyond his original conception of it as only a calculating machine. Babbage himself, rather than diminish Lovelace's contributions to the "science of operations," acknowledged their depth; speaking of the notes that accompany Lovelace's 1842 translation of Italian Luigi Menabrea's (1809–1896) article on the Analytical Engine, Babbage confirms, in contrast to those who believed Lovelace incapable of writing them, that the notes are hers: "The Notes of the Countess of Lovelace extend to about three times the length of the original memoir. Their author has entered fully into almost all the very difficult and abstract questions connected with the subject" (qtd. in Essinger, n.p.).

More than Babbage's confirmation, a review of the notes suggests a level of manual computation and algorithmic thinking that earn Lovelace her contemporary and well-deserved status as the first programmer. Today, Lovelace and Babbage both enjoy a notable, rather than mere footnotable, place in the history of information technology, including Seattle's Ada Developer Academy for training women software engineers, and the feminist journal *ADA: A Journal of Gender, New Media, and Technology*, the internationally celebrated Ada Lovelace Day each October 14, the Charles Babbage Institute in Science and Engineering at the University of Minnesota, and the Charles Babbage Research Center, in Manitoba, Canada. And there is even a Lovelace and Babbage iPad app, based on Sydney Padua's website "2D Goggles" and graphic novel (2015) *The Thrilling Adventures of Lovelace and Babbage*. I discuss in more detail the graphic novel homages to Lovelace in chapter 4. Some co-opting of the Lovelace legacy is more problematic, however. In 2014, the Dear Kate underwear brand debuted its "Ada Collection" to mixed reviews from women IT professionals. The ad campaign features female tech executives of varying body shapes and ethnicities in their underwear, with some rationalizing that it reflects women who are "beautifully real," "successful," and "unique" (Bahadur, 2014), as opposed to objectified in the name of women's economic advancement and feminist empowerment in IT.

Meanwhile, Hedy Lamarr's life has many gendered parallels to Ada's. In the 2004 documentary *Calling Hedy Lamarr*, Lamarr's son Anthony Loder (from her marriage to actor John Loder) enters a wax museum looking for "mom," to discover that damage to her waxed visage led to her removal, destruction, and replacement by a display of Lara Croft, Tomb Raider. As in life, her wax figurine had literally been removed from the pedestal that stardom afforded. At the heart of *Calling Hedy Lamarr* is a son's desire to "know" his complicated mother, conflicted in her roles as woman, wife, and mother, and frequently absent from Anthony and sister Denise's lives to the point that Denise shares her childhood practice of playing with Hedy Lamarr paper dolls as a way to be close to her distant mother, ruining the paper cutouts with her tears. Ironically, Anthony would emerge as a successful telecommunications executive, and Denise would develop a later career as an artist, rendering portraits of the rich and famous, her mother among them. I focus on the most recent Lamarr documentary *Bombshell* in later chapters.

Like Ada, Hedy undoubtedly wanted to escape these traditional gender roles and the circumstances in which she found herself, the escalating crisis in Europe and her Nazi's sympathizing munitions manufacturer first husband, Fritz Mandl (1900–1977). Her status as an Austrian Jew, a fact of which most individuals were unaware throughout her life, made her despise the Nazis and see the rise of Hitler as a threat to Europe and beyond, as she listened and learned from Mandl and his associates about the tools of war, particularly the use of German torpedoes. Soon after escaping from her marriage to Mandl, who both forbade her opportunities to pursue acting and attempted to purchase and destroy every available copy of the scandalous film *Ecstasy*, Hedy already was establishing herself in Hollywood through early films that included *Algiers*, with Charles Boyer, and *Boom Town*, with both Clark Gable and Spencer Tracy. From her safe, increasingly privileged space in Hollywood, Hedy was well aware of the devastation of torpedo warfare evident around the world, as civilian vessels were frequent targets, including the ill-fated *SS City of Benares*, sunk by German torpedoes in 1940 while attempting to carry British citizens, including 90 children, to Canada to escape the Blitz. Despite her desire for freedom, however, Hedy kept entering into marriages. With a newly adopted child of her own, and despite her status as a twice-divorced single parent, she had to do something to fight the war effort.

Still relatively new to Hollywood and not one to frequent the party circuit, Hedy had time on her hands to think and invent. But, not unlike Ada, she wanted an intellectual collaborator. Enter George Antheil. As Rhodes chronicles in *Hedy's Folly* (2011), Antheil's and Hedy's paths were bound to cross, both by virtue of overlapping time in Europe in the 1930s and by virtue of their parallel pursuits in the arts, given George's status as a concert pianist and as a writer, including the prophetic 1940 book *The Shape of War to Come*. Born in Trenton, New Jersey in 1900, Antheil was a diminutive man who stood under five foot five who spent his youth infatuated with European composers such as Igor Stravinsky. His first European trip was sponsored by the wealthy Mary Louise Bok (1876–1970), founder of the Curtis School of Music, in the form of a tour she financed in which Antheil secured a last-minute spot as a replacement pianist. The child of German immigrant parents, Antheil soon found himself in Berlin where he witnessed the desperate measures people undertook to survive in post-war Germany. He also met both his mentor Stravinsky and his future wife, the Hungarian Boski Marcus.

Similar to both Stravinsky and to Charles Babbage, George Antheil had a desire to innovate, in this case to present a distinctly American sound to world audiences in the form of the *Ballet Mécanique*, an original avant-garde score to accompany a short film project conceived by artists Fernand Leger, Dudley Murphy, and Man Ray. A triangulation of the musical, the industrial, and the mechanical, the musical concert was first performed in Paris in 1926, and the cacophony of sound, which included multiple airplane propellers, created such a love-hate reaction in the audience that a riot erupted in the streets. Attempting to recreate the fervor of this debut performance, Antheil introduced the *Ballet Mécanique* at New York's Carnegie Hall in 1927 to a resounding failure by the general audience and professional critics alike, as they were almost literally blown out of their seats. This notoriety found Antheil in the 1930s and 1940s to be an artistic and cultural nomad, writing a crime novel, migrating to Hollywood, and struggling to survive with his wife and young son as he cobbled out a range of jobs that kept them afloat in prewar years.

In light of their shared travels and locations, including 1940s Hollywood where George was writing for *Esquire* and composing movie scores, it is not surprising that Antheil and Lamarr's conversations would turn to the war. Notably, Antheil had experienced the tragedy of war firsthand when his brother Henry was killed in a 1940 plane crash of a Finnish commercial airliner shot down by the Soviet Union for violating airspace blockades. According to Antheil's 1945 biography *Bad Boy of Music*, the conversation about inventing a weapon to stop the Germans began as early as their first private dinner. As with the unlikely partnership between Ada and Babbage, the intellectual collaboration of Lamarr and Antheil defies convention; indeed, as Richard Rhodes puts it in *Hedy's Folly*: "How did an actress and a composer go about inventing a remote-controlled torpedo? What was original about their invention that allowed them to successfully patent it, as they eventually did" (p. 142). Their evolution of "frequency hopping" is attributable to the newfound ability in the 1940s change a radio station frequency via remote control, or perhaps Antheil's understanding of synchronizing player pianos in his earlier *Ballet Mécanique*, and his own work as an inspector of artillery ammunition at a Pennsylvania armory soon after World War I (Rhodes, p. 153). Regardless of the success of the initial patent in 1942, the United States Navy viewed the guided torpedo as largely unfeasible and useless, presuming the technology would be large and bulky, and preferred the antiquated, manual "hit and miss" approach. Their patent expired in 1959, the year of Antheil's death, and the military kept the invention secret for decades as they experimented with spread spectrum technology, never expecting that its use in both government and industry would, by the twenty-first century, become a multi-billion-dollar enterprise, initiated by a woman who would receive too little accolade too late in her own lifetime. And thus, like Ada and Babbage, Hedy and George were ahead of their time.

Most biographies of Lamarr chronicle an internal turmoil, a tension between a presumably glamorous life lived on screen, a life dependent on a physical beauty that gave her little pleasure; as she is often quoted, "Any girl can be glamorous. All you have to do is stand still and look stupid," a vacuous role she had wanted to escape with Mandl but found herself increasingly expected to occupy in Hollywood. Lamarr was not stupid, yet as she aged, she was increasingly anxious to maintain her beauty, given her awareness of its ability to provide a livelihood as her film career ultimately declined. The realities as she aged were harsh ones, leading to the multiple surgeries so common in today's Hollywood, and to a gradual fall from the pedestal that she once ascended in films such as 1942's *Ziegfeld Girl*. She was charged with multiple counts of shoplifting, sued the publisher of her ghost-written publication of a kiss-and-tell biography, *Ecstasy and Me: My Life as a Woman* (1966), that made more of her explicit romantic exploits than her creative genius, and initiated a range of lawsuits that included one against the Corel Corporation for its unauthorized and uncompensated use of a vector illustration of her on the package of its graphic software application CorelDraw (Lettice, 1998), a type of appropriation that is similar to the use of Ada Lovelace to market women's undergarments. Perhaps the graphic artist who used her image, John Corkery, was aware of her technological contributions when he adopted the design that included her once legendary visage; however, it seems another attempt to reduce Lamarr, who ultimately received a lucrative settlement and licensing agreement to live out her final years in relative comfort and relative obscurity, to her superficial status as the most beautiful woman in the world.

Despite their adopted vocations and homes, one the discipline of mathematics over the more creative arts, and another Hollywood over her beloved Austria, both Lovelace and Lamarr wished to be interred closer to their origins. In Ada's case, a later visit to her father's grave led her to her final wish to be buried near him, and in Hedy's case her ashes were scattered by her children in the Viennese woods she never returned to in life. Perhaps if Ada and Hedy had been as known for their innovations in their lifetimes, regardless of adversity, as Stephen Hawking and Alan Turing had been in theirs, and their contributions to scientific and technological inquiry had been better documented and preserved, their stories would alight the screen. Such dramatizations are reserved to contemporary innovators like Steve Jobs, a dramatization I address in chapter 3. Jobs's meteoric rise, fall, and resurrection as Apple co-founder and CEO and early death from pancreatic cancer in 2011 is the subject of multiple books, graphic novels, and films, so much so that these depictions warrant their own Wikipedia page. While Ada and Hedy's life histories are less known, there have been efforts to introduce adolescent girls to their accomplishments, including popular biographer Mary Dodson Wade's 1994 *Ada Lovelace: The Lady and the Computer*. In addition to her biographies on Christopher Columbus, the politician Sam Houston, Olympic champion Wilma Rudolph, and former U.S. Secretary of

State Condoleezza Rice, Dodson Wade's large print biography admittedly does not gloss the brief life of Lady Lovelace. Instead, she balances the assessment that Lovelace's mathematical mind went far beyond the thinking of most of her tutors and mentors to her eventual fall into familial and societal disgrace through unpaid gambling debts, a supposed affair, and an opium addiction, developed in response to the frequent illness and pain culminating in her cancer diagnosis.

Lamarr's story has been chronicled in graphic form in the 2007 *Hedy Lamarr and a Secret Communication System*, written by Trina Robbins, and illustrated by Cynthia Martin and Ann Timmons. Despite the back cover's emphasis on "the story of rich, beautiful actress Hedy Lamarr," and in ways that cloud the collaborative relationship between Hedy and George Antheil, the brief graphic-genred biography acknowledges the double bind in which Lamarr frequently found herself: "People think that because I have a pretty face that I'm stupid. I have to work twice as hard as anyone else to convince people I have a brain in my head" (p. 7). Unlike Dodson Wade's chronicle of Lovelace, there isn't time or space for the scandals that overtook Lamarr's later life and admittedly obscured her technological contributions. Robbins's story is a redemptive one, giving Hedy, and to a lesser extent Antheil, credit for the frequency hopping that has led to cellphones, wireless Internet, satellite systems, and other modern innovations.

CONCLUSION: RE-TELLING THE STORY

As with Jobs, Hawking, and Turing, Ada Lovelace's and Hedy Lamarr's stories are larger than life, combined with the literary history, classic Hollywood, and the propensity toward scandal that makes for big screen melodrama. Yet their stories are largely unknown to the larger culture, all too infrequently produced, distributed, and consumed in the various type of media I've chronicled in this chapter. These stories, and the stories of many other women, nonetheless represent significant technofeminist counter narratives that recover the realities of women's contributions to the rise of information technology. Their stories also challenge the perception that innovation, from Charles Babbage onward, is the result of individual creative genius, and represent invisible partnerships that have historically included women. Both Lovelace and Lamarr are representative of the trajectory Steven Johnson (2014) articulates in *How We Got to Now: Six Innovations that Made the Modern World*. For Johnson, innovation is often all too connected to a single individual, the result of a spark of sudden creative genius as opposed to collaboration that evolves over time and history. It is this trajectory regarding information technology, and women's significant place in it and lived experience of it, that needs to be made more visible in larger portrayals of tech-innovation in popular media genres.

Overall, this book represents a technofeminist historiography that distinguishes the larger cultural, circulatory rhetorics of information technology with the realities of women's lived experience of that culture. In that culture, men are the presumed makers and tellers of the technological progress narrative, or the complexities of the technology itself. In her "Men (Still) Explain Technology to Me" that I reference earlier in this chapter, Audrey Watters repurposes Rebecca Solnit's (2008) original emphasis on what has come to be known as "mansplaining" to chronicle her own experiences as a female educational technology specialist, ultimately concluding that "The problem isn't just that men explain technology to me. It isn't just that a handful of men explain technology to the rest of us. It's that this explanation tends to foreclose questions we might have about the shape of things" (online). In response to this and similar calls, the purpose of this chapter and the ones that follow is to help reshape the cultural narrative of women and technology, a goal that that has political, economic, and social consequences not just for women today but for our understanding of the past, present, and future of technological innovation.

Chapter Two

Distinguishing Rhetoric from Reality in Early Computing Culture

George Antheil died in 1959 never knowing in his lifetime the significance of his technological collaboration with Hedy Lamarr on the digital revolution of the twenty-first century. But by the time of his death, the nascent rhetoric of the computer revolution was a solid presence in the contemporary media of the day. Although the significance of George and Hedy's partnership would not be visible for several decades, they continue to remain much lesser known figures in the popular history of technological innovation. Meanwhile computers such as ENIAC, or Electronic Numerical Integrator and Computer (1946–1955), and IBM's Automatic Sequence Controlled Calculator, known as Harvard's Mark I (1944–1959), dominated both news and entertainment media in post-war America. Yet what would remain secret across media forms was the role of women in the advancement of the "giant brain," as ENIAC was labeled in the news. Indeed, even today, an IBM website overviewing Mark I's history credits Charles Babbage in that "the largest electromechanical calculator ever built and the first automatic digital calculator in the United States" that "brought Babbage's principles of the analytical engine almost to full realization, while adding important new features" (IBM's ASCC Introduction). This IBM archive credits Howard H. Aiken (1900–1973) as the "conceiver" of the Mark I, and thus in this particular technological history, Aiken's collaboration with Grace Hopper, just as Babbage's collaborations with Ada Lovelace, are notably absent. While many information technology histories and biographies have attempted to rectify this depiction of technological innovation as male, such female contributions have been continually marginalized in mainstream media over the last half-century, including the "Women of ENIAC," Frances Betty Snyder Holberton (1917–2001), Jean Jennings Bartik (1929–2011), Kathleen McNulty Mauchly Antonelli (1921–2006), Marlyn Wescoff Meltzer (1922–2008), Ruth Lichterman Teitelbaum (1924–1986), and Frances Bilas Spence (1922–2012), who were for many years unnamed and uncredited for their contributions to the success to the first electronic general computer. Just as Walter Issacson admitted his lack of knowledge of Ada Lovelace, his discussion of the ENIAC in *The Innovators* does inevitably credit its development to John Mauchly (1907–1980) and J. Presper Eckert (1919–1995) "with the help of dozens and engineers and mechanics plus a cadre of women who handled programing duties" (Kindle edition, n.p.). It is important to remember that Issacson's book was published in 2014.

As Judy Wajcman (2004) contends in her book *TechnoFeminism*, "social scientists have increasingly recognized that technological change is itself shaped by the social circumstances within which it takes place" (p. 33). Wajcman and others like Carole Stabile (1994) encourage feminists to "steer a path between technophobia and technophilia . . . to explore the complex ways in which women's everyday lives and technological change interrelate" (p. 6). Although Wajcman focuses on contemporary technologies, she provides a useful historical model in her analysis of the typewriter as a feminized tool in its technical specifications—similar in design to the sewing machine and the player piano, both considered female activities—and in its impact on the rise of women as a working class as emerging discourses about the social acceptability of female presence in male business settings. Similarly, the rise and later decline of women in the early computer revolution of the twentieth century is also tied to both the same cultural assumptions about women's roles in society and the disruption and later reinforcement of those assumptions during and after World War II.

Part of that complex exploration for which Wajcman calls must align the story of women's role in technological innovation with the larger historical and contemporary cultural discourse about technology's alignment with patriarchal power and conceptions of both masculinity and femininity. Such discourses have historically been unchallenged and thus reinscribed in the narrative of technological progress. To establish that alignment, this chapter first chronicles the contributions of and collaborations with women in the evolution of ENIAC, relying on several more recent documentaries marketed to feminist activist groups as well as educators hoping to increase adolescent girls' interests in STEM. Despite this notable collaboration among women, and between women and men, I shall also chronicle how that collaboration was systematically ignored in the media of the day in ways that promoted technological innovation as invention, and programming as lesser, deskilled work, in part because of the reliance in both pre- and post-war era upon women as "computers." That emphasis on individual genius or male partnership is one that captured the media's imagination early on, and thus this chapter compares and contrasts the lived experience of women working on

ENIAC with the portrayal of women as workers in the 1957 film *Desk Set*, one of many of Spencer Tracy's and Katharine Hepburn's later gender-sparring collaborations. Such a comparison also includes Grace Hopper's work on UNIVAC and later MARK I, including her leadership roles and military collaboration, and with her commander and supervisor Howard Aiken, the person who ultimately referred to Hopper as "a good man." Hopper's role as an innovator has long been celebrated both during and after her long life, yet as with Ada Lovelace and Charles Babbage, her partnership with Aiken was one where she, rather than Aiken, could envision the possibilities. While much of this collaboration is addressed in Walter Isaacson's *The Innovators*, it is important to include it within this chapter to further establish the role of women in the history of innovation, including in popular science and biographical treatments such as Isaacson's 2011 biography of Steve Jobs. Thus, as a counterpoint to the historical narrative of these gendered partnerships, chapter 3 includes a contrastive analysis of Jobs and aspects of technological collaboration are portrayed in two popular films about him that bear his name: 2013's *Jobs* and 2015's *Steve Jobs*.

THE PERCEPTION OF WOMEN'S ROLES: NOW AND THEN

I began *Technofeminist Storiographies* in the winter of 2015 at the height of film awards season and the dramatization of science and technology innovators, specifically via *The Imitation Game* and *The Theory of Everything*. By the fall of 2015, Republican candidate and now President Donald J. Trump and his take-no-prisoners approach to public discourse has left no one unscathed, including former Hewlett Packard CEO and Republican presidential contender Carly Fiorina. Paul Solatoroff's 2015 *Rolling Stone* article captures Trump's openly hostile take on many of his competitors, including Fiorina, when her image appeared on the television screen running in the background of his hotel room. As Solatoroff chronicles: "Trump's expression sours in schoolboy disgust as the camera bores in on Fiorina. '*Look* at that face!' he cries. 'Would anyone *vote* for that? Can you imagine that, the face of our next *president*?!' The laughter grows halting and faint behind him. 'I mean, she's a woman, and I'm not s'posedta say bad things, but really, folks, come on. Are we *serious*?'" Although Fiorina was the only candidate to have risen in the polls sufficiently to move from the second tier debate candidate to the first, Trump's comments reduce Fiorina's candidacy and her potential for success not merely to gender but to his assumption of what acceptable gender performance for women should be. As the American public would observe, this strategy was escalated once he and Democratic candidate Hillary Clinton became rivals, with numerous instances that targeted Clinton for not "looking" presidential.

Although Trump later claimed during the September 16, 2015, Republican candidate debates that he was referring to Fiorina's persona, the damage has been done. Fiorina, to her credit, fought back with a powerful video produced by her political action committee "Carly for America" titled "Faces." The video begins with Fiorina speaking to the crowd, "Ladies, look at this face, and look at all of your faces. The face of leadership." As the video continues, women's faces appear on the screen, younger, older, but mostly white, with only two African-American women. Certainly, the Fiorina campaign marshals Trump's sexism and misogyny to outrage its female audience. Equally important, from a feminist perspective, it represents an opportunity—albeit limited by focusing, as Sheryl Sandberg does in *Lean In,* on the experiences of white upper middle-class women—to talk back to those discourses that are mythologized in our culture and that allow Trump's history of violent hate speech against women, their looks, and their presumed roles to resonate with rather than to repulse those who identify with his vitriolic rhetoric. As recently as August 2018, the Fiorina "face" story is back in the news, as the president was criticized for referring to former White House staffer Omarosa Manigault Newman as a "dog," leading to a media chronicle of similar slurs (Bump, 2018).

As Amanda Taub (2015) notes in her discussion of how and why Republican presidential candidates struggled to name a prominent woman to appear on the ten-dollar bill in that September 16 debate, "A big part of this is that there's this disconnect between the traditional gender norms that the GOP still tries to embrace, and the reality of women's lives today" (online). That reality is often obscured in the "face" (pun intended) of an essentialized everywoman, a caretaker of home and hearth, and a stereotype that enabled most of the Republican men on the debate stage to name their mothers, their wives, or other famous caregivers throughout history, and not even American history, as at least one candidate, Ohio Governor John Kasich, named Mother Theresa. Others named Abigail Adams, wife of the second president John Adams, and Clara Barton, founder of the American Red Cross. As expected, Rear Admiral Grace Hopper was not on the Republicans' list of options. For Grace Hopper and the women who were her technological contemporaries pre- and post-World War II, the opportunity to work on top secret initiatives like ENIAC often represented a temporary reprieve from the expectation that they become wives, mothers, or if they did work, teachers. As Martin Gay (2000) chronicles in his discussion of ENIAC programmer Betty Snyder Holberton, her mathematics professor indicated she would better serve her country by staying home to raise children rather than enroll in his course.

Feminist advocates across professions have worked to take on these assumptions about women's roles and women's identities being primarily tied to appearance and to do so across media genres. For example, the 2011 documentary *Miss Representation* is produced and directed by Jennifer Siebel Newsom, a Stanford degreed CEO who has also highlighted the film's theme in a larger non-profit initiative, the Representation Project. Newsom's project makes the compelling case that despite the fact, at the time of production, that women obtain 58 percent of college degrees, and have 85 percent of the purchasing power, they only rank at 33 percent in the numbers of women in the top income countries who have legislative roles, something that, given the inability of the Republican candidates to name women for placement on US currency, clearly establishes a cause/effect relationship. Equally significant is that women have comprised only 17 percent of "all directors, executive producers, writers, cinematographers, and editors working on the 250 domestic top grossing films." As the Miss Representation site identifies the issue, "the media is selling young people the idea that girls' and women's value lies in their youth, beauty, and sexuality and not in their capacity as leaders. Boys learn that their success is tied to dominance, power, and aggression. We must value people as whole human beings, not gendered stereotypes."

Similar to other documentaries on gender equity, there is a strong curricular component designed to reach students from kindergarten to college, yet in this case there is a substantial fee for various forms of licensing for educational groups that make access cost prohibitive for public performances as opposed to individual viewing. This makes the impact a questionable one, particularly given that the video is designed to counter the larger representations in primarily male-produced media that diminish women and reduce them to those traditional roles and identities. Recent social media initiatives sponsored by the Representation Project include media-watch reporting on Facebook, Twitter, and Instagram that as of today even take on, through a series of hashtags: #NotBuyingIt, #MediaWeLike, and #AskherMore, Donald Trump's 2015 blaming of a "young intern" for supposedly mistaken tweets at Iowans who were favoring then competitor and now presidential cabinet member Ben Carson in the polls. As we now know, in Trump's case, the use of Twitter as a bullying tool was just beginning, and included female news reporters, beauty pageant contestants, accusers of sexual misconduct, and political opponents.

In the 2010 PBS documentary, *Top Secret Rosies: The Female Computers in World War II*, the stories of the female programmers of the World War II era are told by the women themselves, women with names, including the ENIAC programmers Jean Jennings Bartik and Marilyn Wescoff Meltzer. As these and other women shared their stories, similar narratives emerged. These shared stories included early affinity for mathematics and science despite limited means to pursue educational opportunities, mentors who encouraged rather than discouraged that affinity, along with the recognition that they wanted alternative career paths that paid more than women-centered professions, including the typical role as a teacher, or they were not ready for the early marriage that was expected of so many young women of the era. Included in the documentary is a large group of women who were recruited by the US military to apply their substantial mathematical skills to improve the precision of ballistic trajectories. Producer Leann Erickson asserts that "Rosie the Riveter made the weapons, but the human computers made them accurate," and yet, their contributions went unnoticed for many decades. Granted, wartime security protocols impacted the covert role of female "computers" working in the United States at the University of Pennsylvania as well in Great Britain at Bletchley Park. And the women themselves maintained oaths of secrecy, not even sharing their work with immediate family members either during or post-War.

While this work was a vital component to the Allied success, such computing and programming tasks became a deskilled form of women's work, an almost clerical task that may have put women in the wartime workplace, but mirrored sweatshop-like conditions. In service of their country, women worked long hours and, with the advent of ENIAC, in oppressive hot conditions. They were the ones actually making the computer work, even though they had no previous access to the machine because of their initial lack of government security clearance. The "Rosies" documentary reveals the immense pride and sense of obligation they felt, similar to their Rosie the Riveter counterparts. Yet it was a role that intentionally went undocumented in the media of the day and, as a result, had substantial consequences both for the published history of technology innovation and for the limited numbers of women pursuing careers in STEM. When ENIAC first went public in 1946 in newsreels and newspaper articles, the collaborative roles of the six women involved went unrecognized. As Issacson chronicles in *The Innovators* (2014), the formal dinner took place at the University of Pennsylvania for military officials, scientists, and university administrators, and just as with the media coverage they were celebrating, the women were nowhere to be found. A February 14, 1946, *New York Times* article reported, "Ceremonies dedicating the machine will be held tomorrow night at a dinner given a group of military and scientific men at the University of Pennsylvania" (Kennedy, 1946).

Most of those wartime contributions are lost. As a result, Erickson's larger goal in making the documentary is to encourage "girls and young women that there is a place for them in hard sciences, mathematics and engineering," made by the women who worked as codebreakers, machine programmers, and ballistics calculators. For Erickson, "Every day that a young woman walks through a door that they opened, the legacy of these women lives on." As computer historian Nathan Ensmenger concludes in the documentary, the significance of the women's programming skills goes unheralded in that it's not merely the computer, but the relationship between the computer and the way it was programmed, a collaborative process for which the women of ENIAC received no public credit and for too long no place in the history of computing in the United States. Ensmenger, speaking of an unnamed museum exhibit of ENIAC, noted that while Eckert and Mauchly were featured prominently, the women, even in the late twentieth century, were labeled simply as the women of ENIAC, unnamed and uncredited for their collaborative work: "We've known for a long time who these women are, they are a crucial part of this story." Even though Ensmenger notes that this occurred several years prior, "their story has still been neglected" in such larger cultural forums, including mainstream media. In his book *The Computer Boys Take Over*, Ensmenger (2010) confirms that the role of programming itself was labeled as "coding" and seen as a form of clerical, even manual labor (p. 35). Ensmenger's history traces the rise of programming as a job duty that created a "labor crisis" in that by the 1960s the demand for programmers far exceeded the supply. Part of that presumed crisis is undoubtedly due to the cultural assumptions about who a programmer was: as an IBM performance report famously put it: "When a programmer is good, he is very, very good. But when he is bad, he is horrid" (qtd. in Ensmenger, p. 19). Good or bad, there is definitely one other thing a programmer was: male.

More recently, there has been an effort to rectify this wrong through similar documentaries and oral histories such as those conducted by the Computer History Museum. NPR reporter Laura Sydell contends that "If your image of a computer programmer is a young man, there's a good reason: It's true. Recently, many big tech companies revealed how few of their female employees worked in programming and technical jobs. Google had some of the highest rates: 17 percent of its technical staff is female." Sydell's report includes interviews with computer science students at Stanford, several women, who acknowledged that while they suspect the first computer programmer was female, they don't know for sure; Sydell concludes that "perhaps knowing that history will show a new generation of women that programming is for girls," as they were in fact among the very first programmers, from Ada Lovelace on.

DRAMATIZING EARLY COMPUTER CULTURE

In her 2001 article "From the *Desk Set* to *The Net*: Women and Computing Technology in Hollywood Films," Carol Colatrella contends that the Tracy-Hepburn classic, in addition to morphing ENIAC into a fictional EMERAC that is personified as the female "Emmy," repositions the role of women in the collaborative reality of ENIAC's success to foreground the Spencer Tracy character, computer genius Richard Sumner, as "the heroic lone male." In this sense, the dramatization of the birth of the computer age in the twentieth-century is not much different from the news reporting of the day. In these depictions, women are subordinate to men, and in the case of the *Desk Set*, the mysterious female machine that is presumed by all to be a workplace replacement for the four women reference checkers, each of them a type of "human computer," at the fictional Federal Broadcasting Network, including Katharine Hepburn's Reference Department Head, "Bunny Watson." When released in the United Kingdom, *Desk Set* was instead titled *His Other Woman*, indicating the competition between the women and the machine as professional rivals, and eventually between Bunny and "Emmy" for the attention and admiration of Richard Sumner. Like her coworkers, Bunny is single, hoping for a commitment from romantic interest Mike Cutler (played by actor Gig Young (IMDB), who notably would be found dead in 1978, the result of an apparent murder-suicide of his fifth wife, thirty-three years his junior), who relies on Bunny mostly for her ability to read and assess financial data for the network, a technical skill set that mirrors the computer she and her colleagues come to fear.

Sumner comes to value Bunny's knowledge as well, impressed with her ability to score highly on an intelligence test that involves more intuitive rather than computational knowledge, leading him to dub her a "rare tropical bird." The romance, which evolves during those times when Cutler is absent, is deferred by the arrival of EMERAC, accompanied by a programmer associate of Sumner's, the ever-efficient Ms. Wariner, played by the character actress Neva Patterson. Ms. Wariner's personality starkly contrasts to the personalities of the four women in the reference department, who without husbands fear for the loss of their jobs given that EMERAC can locate information in minutes that typically took the women days to research. This is not unlike the categorization of ENIAC's benefits when compared to the women computers who would later program her.

Although film historians may focus more on the relationship between Hepburn, whose character ultimately prefers Sumner to Cutler, and Tracy, whose characters ultimately prefers Bunny to EMERAC, Ms. Wariner's character is worthy of closer attention in part because she is the dramatized counterpart of an ENIAC programmer, yet one whose personality is inevitably reduced, like that of the reference department employees, to that of an overly emotional woman who is prone to tears and hysterics when EMERAC malfunctions after several human programming errors that Sumner, as Wariner's supervisor, deems "stupid" and finally "asinine." In contrast to such hysterics, Bunny Watson emerges unrivaled by the mathematical knowledge of either Wariner or EMERAC, who has her own, albeit computational, meltdown, as her counterpart in Payroll prints out pink slips for all employees. Enraged, Sumner indicates that the purpose of EMERAC was never to displace but to help the Reference Department workers be even more efficient, with EMERAC becoming more of a collaborative team player as opposed to a replacement and the women's research roles recognized and valued. When Sumner inputs a query as to whether Bunny should marry Cutler or himself, EMERAC booms an approval, leading to a romantic resolution in which it is unclear whether Bunny will become as subordinate to EMERAC as Ms. Wariner was or become a co-equal partner in the ménage a trois among a man and his computer, though her shift from working woman to wife is similar to the fate of many of the ENIAC women I discuss in this chapter as well.

For Colatrella, this positive resolution represents a "revisionist history" in which the potential enemy is the female machine, as opposed to the men who will ultimate displace women as computers in the rise of computer science as both profession and academic discipline. As she notes, "*Desk Set* represents the battle between machine and human worker as a conflict between masculine devotion to a business economy and the kinder, gentler world of feminine workers who share information and empathy for others as an interim step to marriage and family" (p. 7). Colatrella grounds this conflict not merely to the cultural norms of the time but to the fact that the screenwriters were Henry and Phoebe Ephron, spousal collaborators who wrote Hepburn's character as a substantial competitor to at least the machine, if not the man. A 1957 *New York Times* review of *Desk Set* concurs: "The thought of having Katharine Hepburn as an intellectual competitor is one that should throw fear and trepidation into the coils of any mechanical brain. Miss Hepburn is obviously a woman who is superior to a thinking machine" (Crowther, 1957).

WOMEN'S PROGRAMMED/PROGRAMMING LIVES

The type of collaborative, supportive working environment Colatrella identifies in *Desk Set* was one that did not initially exist for the ENIAC women. As the website for the documentary *The Computers* documents, Kathleen MacNulty indicates that "None of us girls were ever introduced . . . we were just programmers," expected to work in harsh, isolated conditions with little relief. And while camaraderie certainly grew over time, one ENIAC woman eventually moved to the Aberdeen Proving Ground, and these collaborative networks dissipated as the women either moved forward alone, collaborated with male counterparts who in some cases became spouses, or retreated away from the rising professional world of computer science into marriage and family.

Wajcman (1991) asserts that "it was because programming was initially viewed as tedious clerical work of low status that it was assigned to women. As the complex skills and value of programming were increasingly recognized, it came to be considered creative, intellectual, and demanding 'men's work'" (p. 158). Men may have built ENIAC but women were the ones who ensured it was programmed accurately. Regardless, coverage like that of the *New York Times* promoted ENIAC's ability to "compute mathematical problems 1000 times faster than before" and eliminate the need to "accept inferior solutions of their problems, with higher costs and slower progress." By extension, women computers accounted for the "inferior solutions" to problems, and with the success of ENIAC, the six women who originally worked with the team programmed ENIAC sight unseen, relying on schematics of the machine. Despite this feat, their accomplishments were literally ignored for nearly 50 years, not invited to the 50th Anniversary ENIAC Celebration, and their names and their achievements absent from the University of Pennsylvania's historical resources on ENIAC as of this writing, with the exception of a link and general description of the TOP Secret Rosie's documentary as "a documentary of the women employed as "human computers" [*sic*] at Penn during World War II. "Some of them got to see ENIAC before anyone else!" The other specific reference to the ENIAC women occurs in a small uncaptioned thumbnail image in the right sidebar that features two women.

The University of Pennsylvania site also includes a link to 60th anniversary coverage from 2006, a representative article from the popular magazine *Computerworld* featuring a 1989 interview with one of ENIAC's inventors J. Presper Eckert. In Eckert's interview, both he and the article's author Alexander Randall, a professor of computing, do not note specific individuals other than co-inventor John Mauchly; nevertheless, the title of the article, "The Eckert Tapes: Computer Pioneer Says ENIAC Team Couldn't Afford to Fail—and Didn't," acknowledge the importance of collaboration in ENIAC's success, particularly because there were neither precedents nor prototypes. In 2007, the magazine, with a circulation of over 100,000 as of 2012 and a now completely digital format, published a feature on Jean Jennings Bartik, as part of its "Unsung Innovators" section:

> Jean Bartik, born Betty Jean Jennings in rural Missouri in 1924 and educated in a one-room schoolhouse, always dreamed of getting out of the Midwest and having a real adventure in the world. She lived her dream, as it turns out. Bartik was one of six women responsible for programming the ENIAC (Electronic Numerical Integrator and Computer), a giant of a machine charged with calculating bullet trajectories during World War II. Bartik was also the implementer of the first stored-program computer. She helped ENIAC use its function tables to store a programmed instruction set as read-only memory in firmware. Later, she worked on a follow-on computer, UNIVAC, that could store programs in memory. UNIVAC was the first commercially sold computer. (Smith, 2007)

The brief article is likely targeted to their audience of workplace computing professionals and defines Bartik's collaborative role with the other five unnamed women, a programmer, an implementer, and ultimately a helper, subordinate to the men with whom they worked and advanced professionally and personally. Granted, the women of this generation did not have much choice.

As Kathleen McNulty, who would eventually marry the widowed ENIAC co-inventor John Mauchly in 1948, would recall later, "For the big announcement" in February 1946, "the ENIAC Programmers were asked to be hostesses to greet all the big shots and show them around." Their substantial roles as workplace professionals and collaborators were reduced to the cultural assumptions of what women's work was to be. Speaking of her later honeymoon in New York City, McNulty states that "One night after a show, we stopped at a bookstore. When we got back to the hotel room, John said, 'I have a present for you.' When I said, 'What is it?' He said, 'Here it is. It's a cookbook. You are our new cook.' At the time, I thought it was funny. I didn't know how to cook. I'd never cooked." As with her other ENIAC colleagues, McNulty did adapt, cooking for a larger group of extended family that included her five children and one stepchild from Mauchly's first marriage. After Mauchly's death in 1979, his widow was frequently invited to speak about her husband's achievements:

> All the years I gave talks about the ENIAC, I always talked about it as John's story, not my story. Although I mentioned that I had been an ENIAC Programmer, it was just in passing. I told of John's part in the development of ENIAC, about BINAC and UNIVAC, and his companies. I wrote a monthly article about John Mauchly for a little bulletin for some computer club. A member of that club arranged for Jean Bartik, Kathy Jacoby and me to speak at Princeton about our experiences. I think that was the first time that I spoke for myself. (ENIAC Blog, 2004, n.p.)

The ENIAC women and other early programmers were content to return to more traditional roles as wives and mothers, even though, as Erickson contends in *Top Secret Rosies*, their ability to remain in the workforce, largely due to their experiences and mathematical expertise, was more flexible than their blue-collar Rosie the Riveter counterparts. Although all women eventually married (something that was initially grounds for dismissals in the culture of six-day workweeks), Lichterman, McNulty, and Bilas would travel with ENIAC to the Aberdeen Proving Ground Ballistics Research Laboratory in 1947, Bartik Jennings worked in the industry until 1951, only returning to the workforce in 1967, Holberton later worked on developing programming languages with Grace Hopper, and Wescoff Meltzer resigned from the ENIAC project soon after to raise a family. All would be inducted into the Women and Technology Hall of Fame in 1997.

Complementing the *Top Secret Rosies* is the more recent 2014 documentary, *The Computers*, produced by ENIAC Programmer Project founder Kathy Kleiman, an attorney, programmer, and data security auditor, as well as a technology activist who recorded oral histories with four of the six ENIAC women (Jennings Bartik, McNulty Antonelli, Holberton, and Meltzer) in the late 1990s. The documentary's promotional website describes the two-minute film as "an inspirational story that will change stereotypes and throw open doors. It will help students see that technology careers lie within their grasp, and computing professionals know that their field's greatest computing pioneers included women and men!" The website also identifies the documentary's audience as high school and middle school students (an audience similar to that identified by the AAUW in *Tech Savvy: Educating Girls in the New Computer Age*, 2000), along with those who participate in after school computer clubs. Clearly, the battle to revise the male-dominated history through these stories is a longstanding one that continues today, based on labor statistics for tech giants like Google and Facebook that I referenced earlier. As the news articles, websites, and documentaries referenced in this chapter suggest, the stories of these women, despite their collective and individual accolades, is a subordinate one, from the absence of their names in individual articles, to the limited thumbnail representation on historical websites, to the more narrow audiences for the documentary genre as opposed to the larger blockbusters such as *The Imitation Game*, *The Theory of Everything*, and the 2015 dramatization of the life of Steve Jobs in the eponymous film that simultaneously humanizes and demonizes the late Apple cofounder.

Jean Jennings Bartik's *New York Times* obituary (Lohr, 2011) quotes technology historian David Alan Grier: "For years, we celebrated the people who built it, not the people who programmed it." Jennings Bartik's accomplishments could not shield her from the gender and age discrimination she would experience in her later years in the technology industry, when, unable to find work, she began a new career in real estate. Fortunately, technology historians and activists recovered her significant collaborative role in the history of computing in the twentieth century, with a museum in her name, the Jean Jennings Bartik Computing Museum, at the University of Northwest Missouri. As with several of her ENIAC teammates, Jennings Bartik died in a nursing home in 2011, still unknown to the larger public despite noble attempts to better represent the ENIAC programmers' contributions in museum exhibits, documentary film, and oral history genres that usually do not make their way to the big screen. Although these shared histories and partnerships in the history of computing are typically overshadowed by the blockbuster representation of the larger than life male tech innovator as socially dysfunctional genius, from the fictional Richard Sumner in *Desk Set* to the more recent portrayals of Steve Jobs, Grace Hopper's individual and collaborative journey, and how surviving and thriving as a woman in what would become the male-dominated industry we know today, meant being personified as a man.

GRACE HOPPER: QUEEN OR PIRATE

Admittedly, there exist numerous popular and scholarly chronicles of Rear Admiral Grace Murray Hopper's contributions to the history of technology innovation and computer science. In the context of this project, it is perhaps ironic that one such chronicle, Kurt W. Beyer's (2009) *Grace Hopper and the Invention of the Information Age*, has the following back cover blurb:

> A Hollywood biopic about the life of computer pioneer Grace Murray Hopper (1906–1992) would go like this: a young professor abandons the ivy-covered walls of academia to serve her country in the Navy after Pearl Harbor and finds herself on the front lines of the computer revolution. She works hard to succeed in the all-male computer industry, is almost brought down by personal problems but survives them, and ends her career as a celebrated elder stateswoman of computing, a heroine to thousands, hailed as the inventor of computer programming. Throughout Hopper's later years, the popular media told this simplified version of her life story. (Beyer, 2009, backcover)

Compared to Ada Lovelace and Hedy Lamarr, Hopper's life might initially read as less the stuff of dramatic or literary legend. Although Beyer acknowledges that a search on Hopper garners more than a million results and that it was a 1983 *60 Minutes* television news magazine interview that brought her into the public eye, her story, a complex triangulation of historical moment, gender, and generation, is obscured in the larger media narrative of innovation as individual male genius. To be fair, this rhetoric of technological innovation disadvantaged the male "pioneers" of the time as well, from John Mauchly to J. Presper Eckert to Hopper's supervisor and collaborator Howard Aiken. Beyer doesn't identify as either feminist or technofeminist, yet his oral history and biographical methodology counters, as my own efforts in this book attempt to do, the "chasm between rhetoric and reality" (p. 2) that may have mythologized Hopper as "Amazing Grace," "The Queen of Code," or the "Grandmother of Cobol."

Conceding Hopper's evolution as one largely unnoticed by the public, Beyer relies on such oral history to "reconstruct the past" in ways that not only situate Hopper's lived experience within a historical moment but recover that experience as informed by the possibilities and constraints of gender. Beyer's distributed biography of not just Hopper but of individuals and historical events challenges the accepted narrative of the "individual inventor tinkering in a basement or garage for the benefit of society" (p. 19), simultaneously defining Hopper's story as both individual and collaborative, both personal and social, as he documents the cultural and material conditions that may have made Hopper another individual genius in the narrative of technological innovation but almost broke her along the way, including struggles with alcoholism. Beyer moves beyond what has become Hopper's "scripted" public past to the human aspects of Hopper's life that were as complicated as Alan Turing's and Stephen Hawking's, not to mention Ada Lovelace's and Hedy Lamarr's. While Hopper's story is largely confined to the pages of books, the small screen, museums, and websites in her honor, Beyer's reliance on oral history becomes "flesh [that] makes this story so juicy, for we quickly realize that invention as a human endeavor is complex and messy" (p. 17). In the twentieth century that messiness comprised the lives of both Hopper and Lamarr, for while both were arrested, Hopper for drunk and disorderly conduct in 1949 and Lamarr for shoplifting in both 1966 and 1991, far greater events contributed to their parallel life stories. Just as I overviewed Lamarr's personal and political response to World War II in chapter 1, Hopper's journey as a computer programmer has its roots in the War as well, beginning with the Japanese bombing of Pearl Harbor in 1941.

Born eight years earlier than Lamarr in 1906, Grace Brewster Murray's early life was similar to Lamarr's in that she was born into a family of intellectual and class privilege, inspired to pursue mathematics based on her mother's passion for the subject, not unlike Ada Lovelace's mother Anne. Her father supported his daughter's desire to be educated and self-sufficient, and based on her fascination with household clocks, her path toward a career in technology seemed predestined. Hopper's education began at Vassar, with a degree in mathematics and physics, and later a Ph.D. in mathematics from Yale in 1934, the first woman to graduate with an advanced degree from the university. Her life trajectory was both traditional and non-traditional; she married Vincent Hopper in 1930 and took a job as a mathematics instructor at Vassar, working through the ranks of the faculty to associate professor. Part of her professional advancement, as Beyer chronicles, included an emphasis on collaboration on curriculum development and her desire to grow intellectually, auditing courses in the physical and natural sciences and the humanities as well (pp. 26–29). This allowed her to give her courses an interdisciplinary feel in which students could understand the impact of mathematics on the history of ideas and its application to contemporary life.

Nevertheless, Hopper's teaching life at Vassar was becoming a predictable one, one that along with her ENIAC colleagues, she wanted to escape. And just as with Lamarr's strong feelings of patriotism and anti-Nazi sentiment that led to the actress's participation in the American war effort and her own collaboration with George Antheil on the frequency hopping patent, Hopper's journey was equally influenced by her home country's entrance into World War II. In Hopper's case, the shift began as it might for many academics, with a sabbatical, the Vassar Faculty Fellowship, for a year's study with renowned mathematician Richard Courant at New York University, an intellectual reprieve from both personal and professional routines which had become comfortable but boring (Beyer, pp. 29–31). Her intellectual efforts would turn more patriotic by the end of the year, when the Japanese bombed Pearl Harbor on December 7. With her husband, brother, and other male friends joining the war effort, Hopper's mission turned away from what was familiar and comfortable to instead mark a significant change in her personal and professional life.

Wanting to leave Vassar, she negotiated a leave of absence so that she could instead join the navy soon after the passage of the Navy Women's Reserve Act. And despite an inauspicious start to her military career, one that required waivers for her age and her size, her initial training represented a time of focus and concentration similar to her time at Vassar, leading to her position as a battalion commander and being named first in her class. Along the way, Hopper divorced her husband in 1942 and graduated from Midshipmen's School with the rank of lieutenant in 1944. By that time, IBM provided to the Naval Bureau of Ships, who in turn entrusted to Harvard University, what would be known as the world's first computer, the Automatic Controlled Sequence Calculator, or the Mark I (Beyer, p. 37). The creator and commanding officer was Lieutenant Commander Howard Aiken (1900–1973). A Harvard educated physicist, Aiken had seen a demonstration piece of Charles Babbage's Difference Engine, and after a successful partnership with IBM, was allowed to develop the Mark I.

Despite that "partnership," Aiken arranged to receive sole credit for the Mark I in media releases, reinscribing the role of technological innovation as an individual process. Aiken would oversee the Harvard Computation Laboratory more like a ship than a university research lab, a space where the chain of command stressed pressure to get results that would end the war. Hopper's dedication to the war effort, to computing technology, and ultimately to Howard Aiken enabled her to thrive professionally and establish more equal intellectual footing with Aiken, who came to value Hopper and refer to her in later years as "a good man" (Beyer, p. 88). Yet Hopper, as with Lovelace and Lamarr, also became a good collaborator, learning and sharing code with Richard Bloch (1921–2000) and often diffusing the tension of the hierarchical environment with humor and practical jokes (Beyer, pp. 84–85), including the domineering Aiken.

What seems to grab the public's attention is the narrative arc of individual genius or singular heroism, sacrifice in the rise to glory, with a frequent fall from grace or human frailty or failure along the way that leads to society's persecution, redemption, or tolerance as the byproduct of the tragically flawed hero. Such narratives mesh with the larger patriarchal, male-dominated history of progress in athletics, politics, philosophy, athletics, science, letters, and technological innovation. While Hopper's story hasn't yet made its way to the big screen as has Turing's and Jobs's, her story, unlike that of her ENIAC sisters, survived in spite of her gender and her ability to fit into a militaristic and later post-World War II industrial and corporate computing culture. Perhaps this can be explained by Beyer's method of distributed biography, which from a rhetorical perspective, aligns with the classical concept of *Kairos*, that moment in time that requires adapting to that moment, cultural, historical, material, to make an impact. It also aligns with a technofeminist approach, grounded in lived experiences.

Hopper's adapting included those life choices that made her seem to her colleagues and her eventual public more man than woman, as a computer pioneer, as military personnel, and as the lone woman personally and professionally. Beyer includes in Hopper's biography a photo published in the *Christian Science Monitor* in 1946 of both her and Howard Aiken in uniform, examining the piece of Charles Babbage's Difference Engine. Although Hopper is standing and Aiken is seated, she is less behind "her man" than alongside him, engaging in the critical review of the piece of computer history before her (Beyer, p. 131). To Aiken's credit, Hopper's role in the history of computing had been sealed, as Lovelace's had been before her, in part through writing, in this case when Howard Aiken selected, or from her perspective, ordered her to write the manual for the Mark I, completed in 1946. Not surprisingly, *A Manual of Operation for the Automatic Sequence Controlled Calculator* did not credit Hopper as the author, despite her singular role in drafting the fifty-four page manuscript on her off hours. In persuading Hopper to write the book, Aiken reminded her of her military obligation, and the book would be published by members of the computation laboratory, listed by rank, Aiken first, and Hopper third. What made Hopper's manual unique was its attempt to chronicle a history of computing, from Pascal to Babbage and his significant but inevitably failed attempts to secure funding for production of either the Difference Engine or his later Analytical Engine.

Aiken himself saw Babbage's dream realized in the Mark I, though in later interviews both he and Hopper acknowledged Lovelace's role as the first programmer, and the Aiken-Hopper relationship mirrored the Babbage-Lovelace relationship. Aiken and his team at Harvard read Lovelace's published notes, yet those notes were much more than the first algorithm; they were the critical documentation of women's role in the history of digital innovation. Even Aiken, however, would be erased from a competing history of the Mark I when IBM produced its own technical manual, removing individual and collaborative achievements and placing innovation in the hands of the corporation and universities, monolithic and largely anonymous. Such a depiction privileges sponsorship, material and financial, as opposed to individual or collaborative accomplishment, and admittedly, as Charles Babbage's life story suggests, innovation depends on both. Regardless of the competing value systems of IBM, both Aiken's and Hopper's fate were sealed in one version of computing history in general and computer programming in particular, but their respective roles in the hierarchy were vastly different, with Aiken retaining his status as a tenured faculty member and Hopper's three-year contract as a research fellow not renewed.

The pressure to apply their physical and intellectual energy to the war effort impacted Hopper, whose accounts of drinking might be ascribed to the added stressor of surviving and professionally thriving in a hierarchical military culture. And as with the academy, this culture would initially reject her as well, for although Hopper requested to revert to active regular military, she was rejected because of her age. She could have returned to Vassar, where she had started, but chose not to, and unlike the ENIAC pioneers who in some cases balanced work and family or retreated from their careers in computer technology altogether, Hopper persevered, taking yet another risk in transitioning away from the familiar academic environment to the Eckert-Mauchly Computer Corporation (EMCC), where instead of being the lone woman, found herself working with three of the original six ENIAC women—Betty Snyder, Jean Jennings, and Kay McNulty—on UNIVAC, and ultimately mentoring and supervising younger programmers (figure 2.1).

Hopper has credited Snyder for both writing and teaching the code for this latest machine in a culture of collaboration and a less hierarchical management style than she experienced in Aiken's Harvard lab, and Beyer notes that it was a social meeting with Snyder at the home of Association for Computing Machinery co-founder and Mark I colleague Edmund Berkeley (1909–1988) that helped Hopper to decide to join the newer EMCC. Unlike Hopper, Snyder received little credit for her coding and computational flowcharts for the UNIVAC and BINAC digital computers in that her name was never attached to any publication, and her contributions recovered only in part as a result of oral interviews with Hopper in the late 1960s.

It is fitting that Hopper's call for education during her life meshes with her history as an educator of young women in mathematics at Vassar, that the doors that she pushed through be more open and hospitable to future generations. Equally fitting is the recent renaming of Yale University's Calhoun College in Hopper's name in February 2017 (Wong & Svrlug, 2017). The name change is not without controversy, for while Yale administrators and board members wholeheartedly endorse the shift to Hopper, citing her posthumous Presidential Medal of Freedom twenty-four years after her death, National Medal of Technology and other career and military honors as a Yale alumna, her legacy replaces that of the original college namesake, South Carolina Senator and the seventh Vice President John C. Calhoun (1782–1850), a slave owner and defender of both slavery and white supremacy. As a result of the decision to rename the College, which reversed an earlier decision in 2016 to retain Calhoun's name, Hopper was in the news.

Figure 2.1. Grace Hopper at UNIVAC Keyboard, 1960. Courtesy of Smithsonian Archives Center.

Yet in some instances, this coverage is less about a deserving pioneering female technology innovator but more about ousting a historical white supremacist after a period of campus controversy about his continued legacy on the campus.

Such a legacy is connected to this project, for as I have noted, the history of technological innovation has indeed been a white history, leaving the contributions of women of color to the sidelines until very recently, as the book and film versions of Margot Lee Shetterly's (2016) *Hidden Figures* and its portrayal of the overlapping contributions of Katherine Goble Johnson (1918), Mary Jackson (1921–2005), and Dorothy Vaughan (1910–2008) to the US space race while employed at NASA. The debate between what some call an overemphasis on political correctness and what others call bigotry and hate has been a prominent one in the 2016 US presidential election and one that influences even the naming of the building after Hopper. The current controversy surrounding the building name change involves media personality Geraldo Rivera, who resigned as a Calhoun College Associate Fellow, and termed the shift not only politically correct but "lame." In Rivera's Facebook commentary (2017) on his decision to resign the honorary post, he asserts that "once we start messing with history to bend it to contemporary ideals and sensitivities, it might as well be written with disappearing ink."

Such a point presents a suitable foil for the goals of this project overall and my specific discussion of Ada Lovelace, Hedy Lamarr, Frances Betty Snyder Holberton, Jean Jennings Bartik, Kathleen McNulty Mauchly Antonelli, Marlyn Wescoff Meltzer, Ruth Lichterman Teitelbaum, Frances Bilas Spence, and of course, Grace Hopper, in that the sensibilities of their historical moment, the *kairos* of their time, did in fact largely obscure their own contributions to the rise of the academic, military, and industrial computing industries, the rise of computer science as a discipline, and the role of the computer in the larger culture. As a result, even Hopper's legacy, while acknowledged in her lifetime, is nonetheless constrained by the cultural assumptions of gender and technology, leaving Hopper to be an anomaly in her status as a woman among men in her chosen careers of computer programming and military service. In these circles, Hopper's name recognition is not to be taken for granted. For instance, in a February 14, 2017, online newsletter titled *Diverse Military: Higher Education New and Information For and About the Military Community*, the Associate Press feature "Admiral's Name Replaces Calhoun's at Yale" President Peter Salovey was hopeful that the "university community will embrace Grace Hopper and get to know her better."

Recalling the Republican candidate discussion of who should be on the ten-dollar bill and the struggle among male politicians to name a woman of substance, it is fair to say that Hopper's contributions continue to be obscured despite the fact that her rise to prominence as an innovator rival the struggles of Turing, Hawking, and Jobs, but do not align with the larger technological progress narrative, or the US presidential progress narrative for that matter, as male. Yet in many ways, Hopper's journey reflects similar struggles, a complex individual and collaborative, personal and professional, private and public story, including the early alcoholism and threats of suicide (Beyer, 2009) that culminated during her time at EMCC in light of the company's financial instability and her own collaborative role in its success. These complexities may have been left out of the larger narrative that represents the history of technological innovation and the narrative that Hopper privileged in the many speeches and interviews she gave in her later years.

Regardless, by virtue of her ability to not only survive but thrive in adapting to the early computing culture of the twentieth century, Hopper better ensured her role in history. Perhaps, as Yale President Salovey hopes, Hopper's name on a university building will ensure at least to students at Yale, that her story is remembered, certainly for her role in computer programming evolution and for being both an early collaborator on and the public face of Cobol to business and industry. Despite her personal and professional setbacks, Grace Murray Hopper persevered. She worked in the information technology industry and maintained active military status until the age of seventy-nine, six years before her death in 1992 at age eighty-five of natural causes, still working until the end as a consultant to and spokesperson for the Digital Equipment Corporation.

Although Hopper's life is waiting to be dramatized (more on this front in chapter 5), she was immortalized by Google in the 2013 celebratory doodle in honor of her 107th birthday. The Google image is ironically similar to the image of Ms. Wariner from the film *Desk Set* many years before: typing a computational formula into a monolithic machine to get the answer to a question. The early computers, whether human, mechanical, or digital, and the programs written to run them, were perhaps perceived as more scientifically sequestered, their goals and purposes unclear to the larger public. This would account for the lack of popular knowledge not just about the women in computing but the men as well, including Mauchly and Eckert, Howard Aiken, and others. It would be the advent of the personal computer and its impact on daily lives of citizens that would further align modern computing history with the genius innovator narrative.

CONCLUSION: GIVING CREDIT WHERE CREDIT IS DUE

In *The Innovators*, Walter Isaacson (2014) notes that "Like all historical narratives, the story of the innovations that created the digital age has many strands. So what lessons… might be drawn from the tale? First and foremost is that creativity is a collaborative process. Innovation comes from teams more often than from the lightbulb moments of lone geniuses." Another lesson for Issacson was that

> The digital age may seem revolutionary, but it was based on expanding ideas handed down from previous generations. The collaboration was not merely among contemporaries, but also between generations. The best innovators were those who understood the trajectory of technological change and took the baton from innovators who preceded them. Steve Jobs built on the work of the Alan Kay, who built on Doug Engelbart, who build on J.C.R. Licklider and Vannevar Bush (Issacson, Kindle edition, n.p.)

I would argue that in his goal to develop user-centered technological designs that became increasingly portable and accessible in the daily lives of users, Jobs also built on the goals of Grace Hopper, who strongly felt computers would move beyond business and industry to meet the needs of citizen users, and a goal that contributed to her educational advocacy role in her later life.

My goal in this chapter has not been to provide a linear chronological history of the women important to the rise of computer culture in the twentieth century, as that has been taken on through Janet Abbate's important scholarship, as well as Beyer's distributed biography of Hopper and Isaacson's popular discussions of Steve Jobs, among others. Instead, my goal has been to show how those women have contributed to the collaborative as opposed to a genius model of technological innovation that enables the prominence of computers in business and industry, and over time our social fabric. And despite the purported goal of acknowledging women's roles, the historical and contemporary accounts from news media and other popular genres strongly tilts toward that genius myth, however flawed, to explain the rise of technology. Women, figuratively and literally, have been left to the sidebars of such media depictions, and undoubtedly, documentaries and other genres are an important part of the recovering technofeminist storiographies. It is important to question the extent to which an emphasis on these women's lived experiences and the historical records of their vital roles as mathematicians and programmers will change the contemporary dynamics of the information technology industry, which remains male dominated, even in business cultures such as Google and Facebook that purport to be collaborative. Women should enter these fields, as did twentieth-century female computer scientists such as Anita Borg (1949–2003), who founded both the Anita Borg Institute and the Grace Hopper Celebration of Women in Computing to establish community and honor the female legacy within the discipline. Just as with Mauchly, Eckert, Aiken, and Jobs, there are undoubtedly female thinkers from which to draw in the history of technology innovation.

In honor of Women's History Month in 2016, *Mother Jones* published Madison Pauly's (2017) "'I Made that Bitch Famous': A Brief History of Men Getting Credit for Women's Accomplishments," that in addition to a historical timeline that includes Donald Trump claiming responsibility for Lady Gaga's rise to fame and the titular Kanye West, "I Made that Bitch Famous" riff on Taylor Swift, also includes Ada Lovelace, Joan Clarke, the six women of ENIAC, and NASA's Katherine Johnson. For Lovelace, Pauly relies on noted Babbage scholar Bruce Collier, who diminished Ada Lovelace's contribution to to the history of programming by concluding that "it is no exaggeration to say that she was a manic depressive with the most amazing delusions about her own talents, and a rather shallow understanding of Charles Babbage and the Analytical Engine" (qtd. in Morais, 2013). Similarly, Pauly's timeline includes Joan Clarke: "The British government hires Joan Clarke as a codebreaker on Alan Turing's team during World War II. Years later, she is immortalized on screen by Kiera Knightly—but remembered mainly for her doomed engagement with Turing."

For economic, historical, and activist reasons, women should know about their female innovators and role models, from Ada Lovelace, Hedy Lamarr, the six ENIAC programmers, and Grace Hopper. Yet those stories, despite the buildings, and the academic and industrial initiatives to involve women in STEM, remain mostly on the sidelines. Even if they are more visible, they often represent a less diverse storiography that is western, white, and middle class. This project has been a long-term one in that I began in winter of 2015 with one film award season that featured male scientific and technological innovators as flawed geniuses. In winter 2017, another American award season is upon us, and the film *Hidden Figures* is one of the nominees for Best Picture, a testament to the power of women's roles and the injustice of not making them better known to the larger public until the publication and film optioning of Shetterly's 2016 book, along with her advocacy in the form of the Human Computer Project, which she founded in 2013. Given this project's emphasis on collaboration, the roles that Johnson, Vaughan, and Jackson played at NASA are critical ones to acknowledge and celebrate, despite the less direct role they have played in the history of computing compared to a Grace Hopper. Nevertheless, as with the six ENIAC programmers, the contributions of these three women and their many anonymous African-American colleagues who served as "human computers" at the Langley Research Center of the National Advisory Committee for Aeronautics (NACA) at the segregated West Area Computing Unit. Vaughan, for example, left her teaching position in Virginia's segregated school system to work at NACA. With the transition from handing calculations to digital computing Vaughan taught herself FORTRAN and was ultimately named "acting head" until finally promoted to the official supervisor of the unit, where she taught programming to her large team. With NASA's creation in 1958, the organization became desegregated. Compared to the post-World War II context in which the women of ENIAC were encouraged to return to more traditional roles of wife and mother, Vaughan, Johnson, and Jackson were able to persevere in NASA's culture despite larger social values that continued to oppress women and violate the civil rights of African Americans. For example, Katherine Johnson, not unlike Grace Hopper, often found her collaborative report-writing uncredited, yet she became the first women to have her name on a report at the Guidance and Flight Control Division, eventually published 26 co-authored papers, and had NASA's Computational Research Facility named in her honor in 2017.

Mary Jackson's story is equally notable when compared to those of Ada Lovelace and Charles Babbage, Hedy Lamarr and George Antheil, and Grace Hopper and Howard Aiken. Jackson initially worked in Vaughan's unit but moved to NASA's Supersonic Pressure Tunnel to work with aeronautics engineer Kazimierz Czarnecki (1916–2005), who encouraged Jackson to pursue graduate study in mathematics and physics, then offered via University of Virginia evening extension at an all-white high school. Promoted to engineer, her collaboration with Czarnecki evolved, and they co-authored nearly a dozen papers, often with Jackson as first author. Toward the end of her distinguished career at NASA, Jackson moved into administration and mentoring, as a Federal Women's Program Manager in the Office of Equal Opportunity Programs and an Affirmative Action Program Manager. Similar to the recent re-dedication of Calhoun College in honor of Grace Hopper, in 2018, Utah's Salt Lake City School Board voted to change the namesake of Jackson Elementary School to Mary Jackson, rather than President Andrew Jackson, whose slave ownership and policies of displacing Native Americans tarnish his historical legacy (Hicks, 2011). Unlike the cultural myth of Hopper as a tech innovator who succeeded as a result of her rejection of the roles of wife and mother, all three women balanced work and life more traditionally. Johnson was widowed early in her first marriage, which produced three children; she was a single, working parent after her first husband's death in 1956 before remarrying in 1959. Vaughan was married and raising six children during her early years at NACA/NASA, and Jackson was married with two children and served as a Girl Scout Leader. Despite the segregated culture in which they lived and worked, they are the ones who persisted and persevered in their careers, a hopeful model in the face of struggles that women in computing, especially women of color, continue to face today. And all are direct contrasts to the stereotypically techbro computer workforce that positions workers as culturally isolated, reinscribing young white males as presumably more suited to this work context where diverse others, as I document in later chapters, are all too often unwelcome outsiders. The stories of these women, largely recovered by authors and activists such as Shetterly (herself the daughter of a NASA employee), but also through NASA's own dedication to a comprehensive cultural and technological history, are just a few of many that don't make their way to the big screen, a historical exhibit, or a prominent website. This is also due to the segregation of gendered labor roles that distinguished computing from engineering; as Shetterly contends in a 2016 NPR interview:

> It [computing] was "women's work." I mean the engineers were the men and the women were the mathematicians or the computers. The men designed the research and did the manly stuff and the women did the calculations, you know, at the behest of the engineers. And so, I think that it really does have to do with us over the course of time sort of not valuing that work that was done by women, however necessary, as much as we might. And it has taken history to get a perspective on that. (NPR, 2016, n.p.)

Archived at both Iowa State University and NASA, Beverly Golemba's (1994) unpublished manuscript "Human Computers: The Women in Aeronautical Research" provides a personal history of thirteen NACA/NASA women. For Golemba,

> The attitude toward women in science, although somewhat improved, still treats this as an emerging phenomenon, and while young women are presently being encouraged to go into the sciences, the long history of women in science is unknown to them. The young women of today are not aware of the struggles and successes of early women scientists. . . . The dual purpose of this book is to tell their story and to show how these women serve as role models for the young women of today, especially those considering careers in science. (pp. 1–4).

Although Golemba focuses on the need for women in the 1990s to become more aware of early role models who only gained opportunities as a result of a diminished male workforce in wartime, it is clear a quarter-century later that the limited increases in the number of women in computer science fields suggest that the men who dominate those fields need to be made just as aware of the women whose status as "computers" shaped a diverse range of STEM arenas, making both Golemba's and Shetterly's revised storiographies equally necessary then and now. Not unlike the actresses I featured in chapter 1, including Maggie Gyllenhaal and Patricia Arquette in the 2015 awards season, the cast of *Hidden Figures*, notably actress Octavia Spencer, has taken an activist stance on promoting the educational aspects of the film. Spencer made headlines (Romano, 2017) by buying out a theater showtime in Los Angeles, California, so that families unable to afford to see the film could celebrate the legacy of African-American contributions to the space program.

Fortunately, the motion picture industry is not alone in its effort to make visible the lived experiences of women in STEM, as a recent television commercial for General Electric attests. In GE's case, the commercial features the accomplishments of Mildred Millie Dresselhaus (1931–2017), the first woman to win the National Medal of Science in Engineering better known as the "Queen of Carbon," a moniker all too similar to Grace Hopper's status as the "Queen of Code." The ad asks the powerful question, "What if we treated great female scientists like they were stars?" The commercial features the real Millie Dresselhaus on an imagined daytime television program, and her image gracing not only an action figure but even an emoji. Part of GE's stated goal is to create 50–50 gender balance in its technical team by 2020 and, in the words of their Vice President for Accelerated Leadership Lorraine Bolsinger, "Attract, grow and retain a GE technical team that reflects the world in which we live" (Vagianos, 2017). GE's emphasis on a more accurate reflection of lived experience resonates with this chapter and the distinction between rhetorics that obscure women's technological past, present, and future and a more diverse and just reality. While Dresselaus's media immortalization is admittedly more rhetoric than reality, NASA's Katherine Johnson and computer scientist Margaret Hamilton (Abramson, 2017) were awarded the Presidential Medal of Freedom in 2015 and 2016, respectively.

But as the recovery of mathematicians like Katherine Johnson and her *Hidden Figures* colleagues suggest, such women have created parallel histories of the rise of technological and scientific innovation and that there is power in their numbers, if only to make their contributions visible and their voices heard. This goal has led to the development of organizations such as Black Women in Computing, an activist group of support, service, and sisterhood focused on "increasing the number of black women and other underrepresented groups in computing related fields" and creating professional networking opportunities through them. Appropriately, the first formalized meeting of BWIC occurred in 2011 at the Grace Hopper Celebration of Women in Computing, a symbolic testament to Hopper's goals of educating and supporting young people who pursue computing as an educational and/or career path. The organization traces its lineage to Anita Borg, whose work as a systems programmer led to her organizing the online forum "Systers," and before her untimely death from cancer, co-founded both the Grace Hopper Celebration and the Institute for Women in Technology, posthumously renamed the Anita Borg Institute for Women in Technology. Since that time, the organization has also supported a recent annual conference BlackComputeHer. As their website indicates,

> We are a community of tech women interested in changing the narrative around what it means to be a computing scientist. Started by a group of three women in 2016, BlackcomputeHER (pronounced "black computer") seeks to amplify the presence, opportunities, and voices for black women in computing, and tech, more broadly. While gender and race are the most obvious and relatable distinctions among us, those constructs represent a sliver of who we are as people. We join the tech conversation, head on, by offering deep technical knowledge, invaluable resources for professional development, and research expertise in support of *true* tech inclusion. (blackcomputHer.org, n.p.)

Despite the visibility such groups provide to diverse tech workers, it is clear there is much more to do to counteract statistics about the lack of women and people of color working within IT in general and within specific tech giant companies in particular. Blanca Myers 2018 *Wired Magazine* article "Women and Minorities in Tech, By the Numbers" succinctly overviews some of the variables contributing to the lack of women and minorities in the profession. These include the numbers of men pursuing computer science degrees, as opposed to women whose STEM major choices are increasing, but not, ironically, in computer science. Myers also references research studies including observation of IT recruitment interviews, which frequently included sexist jokes and images of men only. Other variables include the access to computer science education; while the AAUW has stressed the subtle discouragement of girls to take interest in technology, these and other students may not have equal opportunity to use computers at home or in the school system. These material conditions align with a technofeminist analysis in that rather than essentializing a lack of inclusion as a lack of interest among a more diverse student or workforce population, it is clear that both attitude about and aptitude for technology is mediated by inequitable socioeconomic conditions and power-knowledge frameworks. Even as the workforce becomes more diverse, there continue to be gaps in pay that are common across industries, with a 16 percent average national pay gap for women in IT, mostly due to the greater disparities in Silicon Valley as opposed to other regions of the United States, and even larger disparities for African-Americans and Hispanics. A 2017 survey by the job website Hired, for example, reports that Black women are offered 79 cents on average to every dollar offered to a white man (Molla, 2017). For that reason, even organizations like the Anita Borg Institute has been subject to critiques about the need to define and prioritize diversity more broadly beyond gender and white women. In response, the institute has made visible efforts to become more transparent and inclusive, releasing annual diversity reports and appointing an African-American woman as CEO, Brenda Darden Wilkerson, former Director of Computer Science and IT Education for Chicago Public Schools and founder of the Computer Science for All Initiative (AnitaB.org, 2017).

Although this chapter has foregrounded representative women's roles in the early history and representation of twentieth-century computing, my next chapter takes on both historical accounts and media depictions of male figures such as Steve Jobs, Bill Gates, and Mark Zuckerberg. There, I assert that the role of women in the histories of both Apple computing and Facebook, along with other Silicon Valley giants, is one that has reinscribed women as all too often the hidden figures in the history of personal computing and social media. These media depictions obscure women's contributions in the same way the women working on ENIAC found themselves erased from that history for nearly a half-century. A consequence has been the limited numbers of women seeking STEM careers, despite the promising future that should have been open to them, something that even philanthropist and computer science graduate Melinda Gates, wife of another tech titan, Bill Gates, is attempting to address. For Gates, who met her Microsoft co-founder husband while employed by the company, the decrease in women earning computer science degrees from 37 percent in the 1980s to 18 percent today is a result of "problems in the education pipeline that push girls away from STEM subjects, adequate role models for young girls interested in tech, and perception problems with male-dominated fields like the game industry" (Statt, 2016). A telling sign of such male domination is the fact that until 2007, a Google search using the words "she invented" would return a query as to whether the user actually meant "he invented" (Wajcman, 2015), revealing that algorithms are not ideologically neutral and both shape and are shaped by larger cultural assumptions about gender, race, and innovation. Safiya Noble's 2018 book *Algorithms of Oppression* proves that the problem is both insidious and ongoing:

> Part of the challenge of understanding algorithmic oppression is to understand that mathematical formulations to drive automated decisions are made by human beings. . . . The people who make these decisions hold all types of values, many of which openly promote racism, sexism and false notions of meritocracy, which is well documented in studies of Silicon Valley and other tech corridors. (Noble, Kindle edition, n.p.)

Gates's and Noble's respective concerns align with Judy Wajcman's (2015) longstanding contention that the "culture of both engineering and computing where the 'masculine workplace culture of passionate virtuosity, typified by hacker-style work, epitomizes a world of mastery, individualism, and nonsensuality'" (p. 181). For Wajcman, it also typifies an emphasis on speed and efficiency, a set of corporate and industrial values that impact the workplace and women's roles in it, even as portrayed in the film *Desk Set*. And, as I will highlight in chapter 3, the real and dramatized culture of the later twentieth and early twenty-first century tech industry has a significant role to play in this process, as does the continued myth of the flawed, rebellious genius. Enter Steve Jobs.

Chapter Three

Bridging the Technological Gender Gap On and Off the Screen

On a wintry Friday morning in January 2018, I attended a strategic planning committee meeting on my campus. As part of that meeting, our university president met with the group to provide its charge; as part of his opening remarks, he shared a number of leadership quotations, including one from Steve Jobs. As I have stressed throughout this book, Jobs's status is legendary across cultures and professions, as his identity in life and death is that of the lone, restless genius able to change the way people live their lives through technology, from the personal desktop and laptop computer to the iPod, phone, pad, TV, and watch we now take for granted today. Not surprisingly, I write this chapter on a MacBook Air while I simultaneously re-watch the 2013 film *Jobs* on my iPhone 8s. Such a scenario, played out in homes, offices, and Starbucks coffee shops around the world, reinforces the role of Jobs and Apple Computers in the history of technological innovation. My purpose in focusing on some of the male "master narratives" of information technology is to show how these stories of innovation have obscured women's contributions in the nineteenth and twentieth centuries, and if we're not careful to counter those narratives, the twenty-first. Thus, this chapter represents a technofeminist reading of the material and cultural conditions of Silicon Valley and their impact on women's continued minority status in the tech industry.

THE STEVE JOBS MYTHOS

The Ashton Kutcher film version begins with Jobs jogging lightly into a 2001 Apple Town Hall meeting, passing by a portrait of Albert Einstein, to launch the iPod to adoring throngs of young people. The film then flashes back to Jobs's humble origins as a college drop-out at Reed College, leaving a one-night stand with a coed he meets on the lawn, only to take all her acid to share with his friend and girlfriend. His intellectual evolution is both spiritual and drug-induced, but doesn't curb his notorious arrogance, as his early role at Atari in the mid-1970s depicts his disdain for designers and co-workers. As he works with his friend, the equally legendary Steve Wozniak, Jobs comments "I just can't work for other people." The Jobs-Wozniak relationship is depicted as a collaborative one, with Jobs able to see the potential in Wozniak's technical abilities to create the first microcomputer, along with Jobs's own ability to communicate the vision that led to the popular demand for what would become the first Apple. The film not only historicizes Jobs's life but the evolution of the Silicon Valley bro culture that begins with Jobs bringing in neighbors and friends to help build computer boards.

In a 2015 interview with the BBC (Kelion, 2015), Apple co-founder Wozniak responds to the Jobs mythos:

> Even while Steve was alive [so was] the myth. He became our hero because we have our iPhones, and he saw the way the world was going to evolve much earlier than others, so we love him. But the mythology, even while he was alive, was extended back to make him that person from day one. [People] try and make him a lot more right about things than he was and ignore what some of the real facts were. (Kelion, n.p.)

Some of those facts are legendary, not so much for their representation of Jobs's leadership but for Jobs's fallibility as both a partner and a friend. These include the paternity denial of daughter Lisa with early girlfriend Chrisann Brennan, who Jobs kicked out of his home once he discovered she was pregnant; his penchant for parking his Mercedes in the mandatory handicapped space at the company headquarters; and his denial of stock shares to his original team, some of whom worked largely for free in his parents' garage to help create the story that intertwines both Jobs and Apple. Nevertheless, Jobs is both an innovator and a guru, displacing the spiritual leaders whose teachings once influenced him, to literally placing Jobs on a grassy knoll as he imparts leadership wisdom to his all-male Macintosh team. The film transitions between the major events leading up to 1996, with Jobs in ascendance after a mostly professional chronicle of highs and lows, and an unexplained inclusion of daughter Lisa in his life. Women within the film are largely peripheral (no technological pun intended), silent stock characters in the background of the male teams, whose collaborations, even that of Wozniak himself, are subsumed to the rhetoric of technological innovation that resides in the lone genius model, though clearly Jobs relied on technical and marketing teams to bring his products and their use value in the larger culture to life.

In this context, recovering the women of Apple is as vital as recovering the women of ENIAC. Indeed, the female members of team Macintosh in the 1980s included graphic designer Susan Kare, whose creation of fonts and icons gave the Macintosh its identity (Kare, 2011), Debi Coleman and Susan Barnes, controllers for the Mac team; PR Executive Andy Cunningham; and the most visible of the Apple women, Joanna Hoffman, whose ability to stand up to Steve Jobs was for these women as inspiring as the creative vision of Jobs himself. A 2015 article (Tibken) titled "Steve Jobs' Legacy Includes the Women He Inspired" stresses the women's gratitude for fostering their creative energy, yet the article betrays the role the women often found themselves playing in managing and appeasing Jobs's fastidious personality and attention to detail in work and life. The interview and panel discussion with the women soon after the release of the 2015 film *Steve Jobs* bears witness to the women who were part of this technological revolution, even as they acknowledge Jobs's verbal abuse. But as with both the Aaron Sorkin dramatization, based in part on Issacson's 2011 biography, and the 2013 *Jobs* version, women are absent, or in the case of the former film, made into a composite character in the form of international marketing executive Joanna Hoffman, depicted by Kate Winslet. As Coleman concludes "Joanna was the one who represented all of us in learning how to stand up to Steve. That's one of the reasons she's a heroine to me" (Tibken, 2015).

Although Hoffman's leadership role is an important one, her singular characterization in the *Steve Jobs* film as the one strong woman denies the collaborative role of the other women who made Apple and its Macintosh a technical, financial, and marketing phenomenon, much in the same way the women of ENIAC were left out of the success stories of the day, as they were depicted in print media such as *The New York Times*. From the 1940s to the 1980s, and as I shall document through films such as the *Social Network* and television series such as HBO's *Silicon Valley*, women's contributions to technological innovation are shadowed by a larger cultural representation of that innovation as male, both individually and collectively, and are as damaging to the future of women in the IT industry and the young girls who aspire to tech careers.

As I discussed in chapter 1, Apple marketing campaigns often created an agonistic, competitive relationship with Microsoft, a perennial enemy that reflected Steve Jobs's desire to identify a foil in both the technology and its creator, Bill Gates. This is most famously depicted in the legendary launch of The Macintosh 1984 campaign, depicting a blond athletic heroine tossing her sledgehammer into a large blue-hued screen from which a dystopic big brother figure proclaims "we shall prevail." The award-winning Super Bowl ad channels George Orwell's ominous prophecy of government policing of both mind and and body to counter that through the introduction of the Macintosh, "1984 will never be like 1984." Yet, as Rebecca Solnit (2014) contends in her article "Poison Apples," "the rise of desktop, laptop, and mobile tools and the social media networks that circulate through them make Apple equally complicit in the surveillance and thought control as the 1984 commercial attempts to resist:

> I want to yell at that liberatory young woman with her sledgehammer: Don't do it! Apple is not different. That industry is going to give rise to innumerable forms of triviality and misogyny, to the concentration of wealth and the dispersal of mental concentration. To suicidal, underpaid Chinese factory workers whose reality must be a lot like that of the shuffling workers in that commercial. If you think a crowd of people staring at one screen is bad, wait until you have created a world in which billions of people stare at their own screens even while walking, driving, eating, in the company of friends and family—all of them eternally elsewhere. (Solnit, n.p.)

The depiction of a strong, androgynous female presence has its roots in Ridley Scott's direction and has often been compared to his dystopic classic *Blade Runner*. The heroine, actor/model Anya Major, was lucky enough to win the role based on her ability to actually lift and hurl the sledgehammer, something other models were not able to do, due to her athletic training as a discus thrower (Wanderings.net). Although the award-winning commercial is legendary in status as "event marketing," the inclusion of more positive and powerful representations of women continued to be more of an anomaly, as campaigns before and after were frequently aligned with white males or past and current male innovators across professions.

Given Jobs's constant emphasis on visionary leadership and innovation, it's not surprising that an early 1980s Apple advertising campaign featured historical male figures such as Benjamin Franklin, Thomas Edison, Henry Ford, and the Wright Brothers. In the Benjamin Franklin ad, the early American statesman is seated in front of a computer at his desk and a kite on the floor, with a large caption in the middle of the page: "What kind of man owns his own computer?" The smaller print includes the statement that "Rather revolutionary, the whole idea of owning your own computer? Not if you're a diplomat, printer, scientist, inventor . . . or kite designer, too. Today, there's Apple Computer. . . . It's a *wise man* [my emphasis] who owns an Apple." This focus on innovation and male dominance continues through the Macintosh's "Think Different" campaign, a visual depiction of diverse male—and sometimes female—characters from varying professions. The campaign designed by Craig Tanimoto features portraits of Gandhi, Muhammed Ali, Einstein, Jim Henson, Picasso, Martin Luther King, Bob Dylan, even Jobs and Wozniak, along with fewer notable women: Amelia Earhart, Martha Graham, Eleanor Roosevelt, Maria Callas, and Jane Goodall. The "Think Different" campaign coincided with Jobs's return to Apple at its lowest point in 1997, near bankruptcy, and evolved in part as a response to IBM's "Think IBM" campaign.

The campaign also included a commercial titled "The Crazy Ones," voiced by actor Richard Dreyfus:

> Here's to the crazy ones. The misfits. The rebels. The trouble-makers. The round pegs in the square holes. The ones who see things differently. They're not fond of rules, and they have no respect for the status-quo. You can quote them, disagree with them, glorify, or vilify them. But the only thing you can't do is ignore them. Because they change things. They push the human race forward. And while some may see them as the crazy ones, we see genius. Because the people who are crazy enough to think they can change the world, are the ones who do. (Siltanen, 2013)

Originally, Jobs himself recorded the voiceover, initially concerned that "putting the Apple logo up there with all these geniuses will get me skewered by the press," a concern that he soon abandoned. The 2013 film concludes with Ashton Kutcher's Jobs recording the mantra that came to culturally represent Jobs as much as the historical icons that comprise the Think Different campaign. *Guardian* reporter Steve Rose (2015) concludes "That Think Different ad also speaks volumes about the type of company in which Jobs expected history to place him, and so far history has obliged." Rose's article includes an interview with *Steve Jobs* director Danny Boyle about his own 2015 biopic. Speaking of Jobs's legacy, Boyle reflects "We're all partly responsible for that. We all bought into it, especially in America. He's a businessman really, but in America it's part of the myth of the frontier: the pioneer. One man forging ahead, breaking through any barrier."

Boyle's 2015 film is as complicated as Jobs himself. The film begins with a 1964 prediction by Arthur C. Clarke (BBC Horizon) explaining the personal computing internet culture of 2001 to a journalist as the reporter appears with his son, noting that the child will be able to live his everyday life through the computer, from bank statements to theater tickets. The reporter asks what the social world will be like in such a computer-dependent life; Clarke's response focuses on the opportunity for *businessmen* [my emphasis] to live life remotely away from the city, interacting with other human beings and conducting business through the computer itself. Clarke focuses on the ability of men to conduct business internationally in a number of segments recorded in the 1960s, able to communicate via satellite, Bluetooth, and wireless technologies attributed to the original frequency hopping system of Hedy Lamarr and George Antheil. Yet, for Clarke, the past, present, and future of technology use and innovation was male, making the work and home life of the businessman easier. Perhaps director Danny Boyle and screenwriter Aaron Sorkin sought to establish Jobs's fulfillment of Clarke's prophecy. But the clip also reinforces the masculinity of computer culture, equally reinforced by the opening scene of the film, which begins with Jobs, Joanna Hoffman, and a technician sparring about the ability to get the Macintosh to say "Hello." While Hoffman cautions Jobs to control his temper as a reporter from *GQ* magazine is in earshot, the mother of his child, Lisa, sits in the audience with the young girl. Lisa, who believes the computer is named after her, is visibly confused and hurt by Jobs's assertion it is not, and that the naming is a coincidence based on the acronym for "locally integrated software architecture."

As in any biopic, fact and fiction are blurred; purported to be fiction is Jobs's softening toward Lisa based on her interest in the computer and her ability to use the paint tool to draw an abstract. What makes this version of Jobs's storiography compelling is Lisa's role in the film and throughout his life as a touchstone for his flawed character, the traits that allow him to deny paternity, deny meaningful financial and emotional support, withdrawing it when displeased. These are traits that extend to his treatment of others in the history of Apple, and something, to its credit, not glossed in the film. Neither glossed was the decision on the part of *Time Magazine*, presumably courting Jobs for its then coveted "Man of the Year" cover (it became "Person of the Year" in 1999), only to find himself displaced by the designation of the computer itself as the "Machine of the Year." The resulting cover story on Jobs became one that featured his denial of paternity and claim that 28 percent of the American male population could have fathered Lisa (Dockterman, 2015). Undoubtedly, part of film's narrative power is due to the involvement of Lisa Brennan Jobs. Aaron Sorkin has referred to Lisa as the "heroine" of the film and someone with whom Sorkin consulted, as was Steve Wozniak, whose pleas within the film to have Jobs credit the original Apple II team represent a collaboration rendered invisible in pursuit of the larger narrative of Jobs as a visionary tech titan. Throughout this project, I have focused on the extent to which this narrative of individual masculine genius has ignored the historical contributions of women and limited the future of women and girls in the computing industry. However, this narrative clearly has had an impact and perception on the collaborations among men as well.

In his memo to Apple employees upon Jobs's death from pancreatic cancer in 2011, CEO Tim Cook wrote:

> Apple has lost a visionary and creative genius, and the world has lost an amazing human being. Those of us who have been fortunate enough to know and work with Steve have lost a dear friend and an inspiring mentor. Steve leaves behind a company that only he could have built, and his spirit will forever be the foundation of Apple. (Cook, 2011, n.p.)

The memo is a suitable tribute that acknowledges Jobs's leadership role and technological vision. Nevertheless, the building of Apple as a company and the technologies we have come to take for granted are the result of many collaborations among men and women, including Bill Gates and Microsoft, Jobs's frequent foil in the early days of Apple. Notably, when it came Bill Gates's time to appear on the *Time*'s Person of the Year Cover in 2005, it was under the heading of "The Good Samaritans" (including U2's Bono) and co-featured his wife Melinda for their development of the Bill and Melinda Gates Foundation. As Rose (2015) concludes, "if Jobs's life has been mythologised, it is perhaps as much a product of our own limited narrative templates as his own self-promotion." That narrative template is one of a singular hero's journey, an American Horatio Alger story with enemies and obstacles along the way worthy of dramatization.

Such an antagonistic relationship, notably between Jobs and Gates, is evident in one of the earliest biopics, *Pirates of Silicon Valley*. Starring Noah Wyle as Jobs and Anthony Michael Hall as Gates, the DVD cover for the made for TV movie reads "The true story of how Bill Gates and Steve Jobs changed the world." History often depicts the evolution of Apple and Microsoft as individual journeys of genius, but in truth both Gates and Jobs were dependent on each other in those early days, with the Microsoft team developing graphical versions of Excel and Word for the Mac II, which ultimately led Gates and Microsoft to develop its own Windows interface. The rivalry depicted in the film culminates in the famous and true exchange between Jobs and Gates when Jobs discovers Gates's plans for Windows and accuses Gates of stealing the idea of a graphical interface, which Jobs himself adopted from Xerox Parc. Inevitably, for both the Apple and Microsoft teams, the film depicts technological innovation and collaboration as something that occurs in dorm rooms, garages, and cramped offices, and the corporations and power-brokering that markets and distributes that innovation as inherently male. As with Grace Hopper before them, Jobs and his team viewed themselves as "pirates," a small group of select rogues opposed to a bureaucracy that Apple as a whole was coming to represent. That pirate mentality may have had consequences for Jobs's ultimate legacy, especially when compared to Gates, for as Jobs gathered the spoils of the PC wars and the shift to mobile and tablet media and communication tools, he chose to channel his energies, particularly after his return to Apple in 1997, into building the empire. Gates, on the other hand, began to shift his attentions, in collaboration with his wife, to a philanthropic model that as some have suggested will ensure that his legacy as a humanitarian outstrips his role as either a tech titan or a pirate.

My point in this discussion is not to argue that Steve Jobs was a despicable human being or less of a tech innovator than the culture makes him out to be. Rather, it is to show how, as Danny Boyle has suggested, the mythology of technological pioneer allows us to reinscribe a "narrative template" that leaves little role for women and creates a Silicon Valley culture that justifies their exclusion. In Walter Issacson's biography, Jobs is quoted as criticizing Gates for his lack of imagination: "Bill is basically unimaginative and has never invented anything, which is why I think he's more comfortable now in philanthropy than technology. He just shamelessly ripped off other people's ideas." In an interview soon after Jobs's death, Gates dismissed Jobs's criticism as "understandable" and the larger myth of a unique individual tech genius that transcends the need for a traditional education when journalist Christiane Amanpour (2011) notes the similarity of Gates, Jobs, Twitter's Jack Dorsey, and Facebook's Mark Zuckerberg's status as billionaire college dropouts. Gates disagrees with Amanpour's suggestion that America needs different paths toward innovation and success that these entrepreneurs' lifestories represent. Referencing Jobs's own journey, Gates concludes:

> I think there are some people who will make their own route. And, you know, those people don't need some guy to make a route for them. I mean I don't think somebody is saying, "Okay, you go up to Reed College, you drop out after six months, you—you take various drugs for a while, you go to India." I don't think that's something we need some well-defined route for that, you know, one out of a million type person. For most people, being able to do mathematics, being able to read, have the jobs skills that would let you be a nurse, the job skills that would let you be a policeman, a teacher, you know, these are great things. And our capacity for doing that needs to expand. (Amanpour, 2011, n.p.)

Although Gates assesses Jobs in ways that reinforce his pioneering status, he resists the idea that such individuals are somehow better for society than the eradication of poverty and the training for careers that his more current emphasis on literacy education privileges. In this way, Gates has much in common with Grace Hopper's call for education, though in Hopper's case, it is her legacy of training young people to program. Gates ultimately disrupts the narrative template even in his own life, which may explain why there has not been the same level of effort to dramatize his life as the founder of Microsoft who transitioned from tech entrepreneur to collaborative philanthropist.

Similarly, the rise of women tech CEOs, including former eBay and Hewlett Packard CEO Meg Whitman, is a life story less chronicled. Whitman, who completed degrees at Princeton and Harvard and was deemed by *The Financial Times* to be one of fifty faces that shaped the decade 2000-2010, ran an unsuccessful political bid for governor, and even shoved and shouted expletives at an eBay employee to boot in 2007 (Mehta, 2010), leading to a $200,000 settlement for that individual. Perhaps Whitman's story doesn't fit the iconographic mold of a Steve Jobs or even a more recent female *The Iron Lady*, the political rise of the late British Prime Minister Margaret Thatcher. Admittedly, Whitman's accomplishments are more entrepreneurial than technological, as we see with Yahoo's CEO Marissa Mayer. As an engineer, Mayer rose through the ranks at Google from programmer to her role as Vice President of Location and Local Services, only to have a male executive promoted directly above her. Many hoped Mayer as a former programmer would be a stronger advocate for educating women and girls for careers in computer science. Instead, she argued that "I'm not a girl at Google. I am a Geek at Google. I am much less worried about adjusting the percentage than about growing the overall pie. . . . We are not producing enough men or women who know how to program" (qtd. in Rosin, 2012). Roxane Gay (2014) discusses Mayer's public rejection of the label feminism as a negative word in a 2012 interview, implicitly critiquing the CEO despite her own insistence that the term *feminism* just like the term *woman* must not be essentialized: "For Mayer, *feminist* is associated with militancy and preconceived notions. Feminism is negative, and despite the feminist strides she has made through her career at Google and now at Yahoo, she'd prefer to eschew the label for the sake of so-called positive energy" (p. 307).

Mayer, the now former Yahoo CEO, who resigned in 2017 after the closer of a merger acquisition by Verizon, and who once claimed the gender of a tech CEO was not important—that tech should be "gender neutral"—acknowledges the double standard facing women, noting the coverage of former presidential candidate Hillary Clinton's appearance and wardrobe, and the ultimate coverage of Mayer herself.

Steve Jobs may have been a "micromanager," but was seen by many as "genius obsessed with details" (Peck, 2016), whereas Mayer has been judged for her appearance, her pregnancies and how they allowed her to keep her position, her decision to pose in a fashion magazine, her childcare choices and the extent to which she could have been a better model for female employees. In addition, a lawsuit was filed against Mayer by a former male employee for his firing, based on a perceived shift to a primarily female editorial team at the tech company. Mayer may have "leaned in" just as Sheryl Sandberg has advocated and has walked away from Yahoo millions richer than when she started, but there has been a double standard for women as opposed to men in an industry where the labor data suggest that over the last five years, both Yahoo and Google's gender numbers are approximately 30 percent female. Perhaps Mayer aimed for gender neutrality because she wanted to escape potential gendered commentary on her leadership roles at Google and Yahoo. Sadly, the opposite occurred, perhaps because we still cannot disrupt the narrative of the male innovator as icon, a position that limits the potentials of both men and women as tech workers and potential IT executives.

Regardless of gender, we also cannot disrupt the narrative of the white executive, and even as I overview representative women CEOs, it is important to complicate the story by considering intersectional politics for women of color in such positions, though as the data suggest, these numbers are low across business and industry. A notable instance involves the gender discrimination lawsuit by Ellen Pao against the venture capital firm Kleiner Perkins after being denied a promotion that she contended that men with comparable accomplishments and credentials were granted, along with claims of retaliation by a co-worker with whom she had had a romantic relationship and was ultimately in a supervisory position over her. Since her lawsuit, the increase in gender discrimination litigation has been dubbed the "Pao effect," while some have expressed concern that the fear of such legal action may lead to Silicon Valley hiring fewer women. After a tumultuous tenure as interim CEO at Reddit where she was influential in banning "revenge porn" and other forms of online harassment and hate speech against many diverse groups, leading to severe backlash by users, coupled with controversial decisions involving termination of other employees and policies about salary negotiation (supposedly in order to benefit women, who are often disadvantaged in such dialogues, as earlier comments I quoted by Microsoft CEO Satya Nadella attest), Pao resigned. Pao, the daughter of Chinese immigrants, lost her case and has spoken compellingly of the personal and professional consequences of "leaning in" in "This is How Sexism Works in Silicon Valley. My Lawsuit Failed. Others won't," her 2017 book *Reset: My Fight for Inclusion and Lasting Change*. The book concludes with her collaborative work on Project Include, "a non-profit that uses data and advocacy to accelerate diversity and inclusion solutions in the tech industry" (projectinclude.org). Pao may have been burned by "leaning in," but it's clear through her more activist efforts that she is attempting to talk back and advocate for inclusiveness, comprehensiveness, and accountability in the tech industry and beyond.

THE SOCIAL NETWORK

The 2011 film *The Social Network* begins with a bar scene between Mark Zuckerberg and his current girlfriend in which Jesse Eisenberg's characterization of Zuckerberg reflects an arrogant, elitist, and insensitive date as he obsesses about his getting into one of the all-male prestigious finals clubs at Harvard. By the end of the first scene, Zuckerberg's girlfriend breaks up with him after he denigrates her status as a student at Boston University and implies their access to the bar they're in is based on her past sexual relationship with the doorman. Zuckerberg's misogynist treatment of his date by posting hateful anti-Semitic and sexist comments about her to his blog foreshadows the common doxing of women in #gamergate and the revenge porn sites that have emerged as users post "wins" of desired victims. The fictional Zuckerberg, in a continued rage, uses his technological prowess to hack the "Facebooks" of seven residence halls to create a "Facemash," a site in which the faces of female undergraduates are paired on a page for male viewers to pick the "hotter" version.

Aaron Sorkin's screenplay, based on Ben Mezrich's 2009 book *The Accidental Billionaires: The Founding of Facebook, A Tale of Sex, Money, Genius, and Betrayal*, is admittedly a blur of fact and fiction. The character of Zuckerberg's fictional girlfriend Erica Albright is exactly that, fiction, as he was likely already involved with then girlfriend, now wife Priscilla Chan, by 2003. Yet the dorm room, private club culture of both Harvard and inevitably Silicon Valley is closer to the truth. This exclusionary boys club as represented in the film mostly reduces women to social climbers hoping to enter the prestigious club set, yet Rashida Jones's depiction of attorney Marilyn Delpy closely resembles the moral conscience role that Kate Winslet will ultimately assume in her performance of Joanna Hoffman in the later Sorkin screenplay for *Steve Jobs*.

Undoubtedly, these films are parallel in their focus on the complication of male friendships and collaborations, with Delpy encouraging Zuckerberg to settle lawsuits against fellow classmates Cameron and Tyler Winklevoss for violating the oral commitment to work on their HarvardConnection site and use the time and source code to create Facebook instead, a presumed theft of intellectual property that ultimately garnered the twins a multimillion-dollar cash and stock settlement. And just in the way that the original group in Steve Jobs's parents' garage are largely, save Steve Wozniak, lost to the larger cultural narrative of tech innovation, so too are the original Harvard dorm-room partners Eduardo Saverin, Andrew McCollum, Dustin Moskovitz, and Chris Hughes. Toward the end of *The Social Network*, Marilyn Delpy tells the fictional Zuckerberg that "you're not an asshole, Mark; you're just trying so hard to be." In a moment of redemption, the film ends with Zuckerberg friending the fictional Erica Albright, earnestly awaiting a response that doesn't come. The real Mark Zuckerberg was named *Time*'s 2010 Person of the Year for "connecting more than half a billion people and mapping the social relations among them, for creating a new system of exchanging information and for changing how we live our lives" (Grossman, 2010). The story also featured a video segment titled "Working at Facebook: A Day with the Profile Team," a group of all male engineers, designers, and data experts. Undoubtedly, the segment represents the collaboration among this team, an important counterpoint to models of individual innovation. Yet the small group of white and Asian male employees also represents a microcosm of the company itself, and as the video crew pans various other spaces at Facebook headquarters there are very few women. As the video journalist concludes, "Just another day behind the code at Facebook."

As with Bill Gates, Zuckerberg has sought to move beyond his status as a Silicon Valley giant, signing Gates's and Warren Buffet's philanthropic Giving Pledge. As the fourth richest person until late July 2018, he and his wife Priscilla have founded Chan Zuckerberg Initiative, which will devote 99 percent of their Facebook shares valued at over 60 billion dollars to align science, technology, and engineering research and education with social justice and world health challenges. Zuckerberg's status as a billionaire, though secure, took a major hit in July 2018 when Facebook stock dropped 19 percent; Zuckerberg's loss was over $15 billion in a single day. My focus on Jobs, Gates, and Zuckerberg has been to chronicle the iconic role of the entrepreneurial, often unscrupulous tech genius, and the extent to which that stock character as represented in popular media is one that eschews the actual collaborative relationships that enabled such technological innovation and its resulting impact on twentieth- and twenty-first-century society. And as I have stressed throughout this project, this characterization also obscures the documented role of women's historical and contemporary contributions to that innovation and to the tech industry as a whole.

THE CHALLENGES OF LEANING IN

As I note in my introduction, the percentages of women in executive and other leadership-level positions within Silicon Valley are unsurprisingly low. These data document that Silicon Valley is a microcosm of the larger lack of opportunity for women to lead. While the law firm of Fenwick and West's "2016 Report on Gender Diversity in Silicon Valley" acknowledges a cultural ecology that normalizes women's lesser role at these and other levels, their rationale also includes choices that put the burden on women themselves for this lack of gender diversity, noting differences in areas of education, career field or industry selection, risk-taking, perseverance, and career interruption. Ironically, the firm concludes its report by including its own workplace diversity initiatives to better support women attorneys. Fenwick and West's data, along with the earlier statistics I cite, clearly reflect the result of the cultural representation of Silicon Valley as masculine with little room for women to tell their story and shape the narrative unless they become one of the few who break through the technological glass ceiling, as we see with Whitman and Mayer. Such is the case with Facebook's Sheryl Sandberg. On a personal note (Beck, Blair, and Gronowski, 2015), I first read Sheryl Sandberg's (2013) *Lean In: Women, Work, and the Will to Lead* on a cross-country flight from Detroit to Reno. Heading to my annual reunion with my four best friends from high school, I had been told I should read the book as someone who was perceived to be a woman who had "leaned in" throughout my career. As an academic administrator about to head into my ninth and final year in this major leadership role, Sandberg's book resonated not just with IT culture but also with academic culture. Squirming in my seat at 36,000 feet, I began to question "Lean in? How about Lean Back?" Although Sandberg's story represents an individual journey, it is in many ways reflective of earlier generations of women speaking about their experiences as programmers, business owners, and STEM professionals in Janet Abbate's (2012) *Recoding Gender*.

Sandberg's book represents the personal journeys of many women colleagues and friends, not just her own. Throughout *Lean In*, Sandberg chronicled her emerging role as feminist champion, sharing both the personal struggles in maintaining work–life balance that she has experienced as the one of the few women chief operating officers in the informational technology industry, maintaining her marriage (she became a widow in 2015) and her role as a mother of two. For herself and for other women in IT, Sandberg acknowledged the various double binds, including that many women lean back because they presume that something will give, either work life or home life, that will make them not able to assume leadership responsibility. In this process, Sandberg is honest about the lack of support structures for working women, tying that to the presumption that the average worker at a Google, a Yahoo, or a Facebook is male and would not need amenities such as a parking spot closer to the building during pregnancy, something she herself only realized when she became pregnant for the first time.

Sandberg also documented the various attitudes that harden the glass ceiling for women in a range of industries, but especially IT. She begins *Lean In* with the overwhelming negative statistics surrounding women's labor and leadership in IT and her lament that there are simply not enough women in power, something she became increasing conscious of as, increasing in rank, she found herself the only woman in the room. This is clearly due to a reinforcement of gender assumptions that often put women in the position of doing more work, of helping more, without recognition or reward because this is supposedly a traditional role that women want to assume because of their "natural" instinct to help and collaborate. Sandberg admitted that failure to assume these roles in favor of a more assertive self-promotion of skills and abilities can lead to dislike and distrust of women employees and thus hinder rather than help their advancement.

Certainly, Sandberg is not alone in this concern about the limited numbers of women, from the boardroom to the developer to the classroom, women in STEM, and in particular "T" for the technology portion of that now ubiquitous acronym that defines the knowledge gap between the United States and other cultures. No one can deny Sandberg's status as an influential leader in IT and her authority to speak about gender and labor dynamics, given her status as the fourth most powerful woman in the world in the recent 2017 *Forbes* poll of the top 25 female leaders, and with her own profile in the comic book series *Female Force*, published by Bluewater Productions (Frizzell, 2015).

But how do women become such a force in the IT industry and beyond? For Sandberg, the answer is partly connected to opportunities for more focused but genuine mentoring and sponsorship of women, something Sandberg credited for her own success. In her chapter, "Are You My Mentor?" Sandberg emphasized the need to shift perspective from "Get a mentor and you will excel" to "Excel and you will get a mentor" (p. 68). This moves the mentoring relationship from the appearance of handholding to a more reciprocal relationship where mentors benefit from the commitment and the knowledge of mentees. Regardless, Sandberg acknowledged the difficulties of establishing mentoring relationships that provide equal opportunities to men and women, both in terms of who gets mentored versus who gets ignored, and who is expected to mentor women, invariably other women, which can lead to yet another layer of invisible labor common to women's working lives. Given the male majority in the tech industry, women are in a double bind, greatly in need of mentors but with very few role models. On the surface, it may seem that the unique stories of women in information technology, including Sandberg's, are not necessarily representative of those of other women, lest we reinscribe the structures that continue to dominate the IT industry and a status quo mentality that positions women and people of color in companies like Facebook and Google as perpetual minorities. As Elissa Shevinsky concludes in her introduction to her 2015 edited collection *Lean Out: The Struggle for Gender Equality in Tech and Start-Up Culture*, "The success of men like Apple co-founder Steve Jobs and Facebook CEO Mark Zuckerberg created the perception that ideal founders of companies look just like they did—young, white, male, and socially awkward." As with the "tableflipclub" and "About Feminism" groups I discuss in chapter 1, Shevinsky describes this collection as a feminist/intersectional manifesto. Manifesto or not, Shevinsky's collection aligns with my goals in her assertion that "The movies and history books and hiring practices at big tech companies may reinforce the idea that young white male nerds have a natural affinity with computers and with code. But the truth is that women—and women who defied their assigned gender roles at great cost—have just as rightful a place among the luminaries of Silicon Valley."

A January 29, 2018, MSNBC Special Presentation *Revolution: Google and YouTube Changing the World* features the CEOs of both companies, the former, Sundar Pichai, a man of color, and the latter, Susan Wojcicki, a white woman, both acknowledging the cultural assumptions that limit women's and girls' educational and career opportunities. In her co-interview with MSNBC's Ari Melber, *Recode*'s Kara Swisher notes that while some refer to

Silicon Valley as a meritocracy that allows the best and brightest, as Sandberg's *Lean In* narrative suggests, to rise to the top, it is in fact a "mirrortocracy" where white guys look at each other and hire each other. For Wojcicki, a larger part of the dilemma is that computer science and tech as an industry is perceived as "geeky" and thus the educational pipeline is narrowed by the stereotype.

But as Swisher contends, sexual harassment within companies like Uber and pay inequities at Google foster an overall hostile environment that marginalizes women, even as Pichai calls for increasing representation of women in tech industry to speak back to the actions of the James Damores of Silicon Valley. Damore, the software engineer who wrote the manifesto in 2017 suggesting that women weren't suited for careers in tech, was fired for suggesting that women were into feelings and not into ideas, an argument I address more specifically in chapter 4 as part of the "bro culture" that excludes the very women these tech giants claim to want to recruit. Wojcicki claims to have expressed hurt at the depiction of women tech workers and its impact on her own children's perception of women's capabilities and their ability to be successful in a particular career path, as well her frustration at her own attempts to recruit women to an industry where women account for approximately 20 percent of the workforce at companies such as Google.

SILICON VALLEY: ART IMITATES LIFE?

Art imitates life or vice versa in HBO's *Silicon Valley*, the story of a group of engineers and programmers, who look remarkably similar to the group featured in the *Time Magazine* Facebook feature. The show is created by Mike Judge of *Beavis and Butt-head* and *King of the Hill* fame. The cover image for the show includes the five-member male team of the fictional start-up Pied Piper, dressed in the proverbial Steve Jobs black turtleneck, and includes the subtitle "where everyone wants to be an icon." The opening episode in 2014 begins at a lavish cocktail party at the home of a successful start-up, with members of the group, including the main character Richard, contemplating the wealth of guests and hosts. As the group comments on the gender divisions and the typical challenge of talking to the women at the party, one opines "They don't have to talk to girls. This house talks to girls." This sexist presumption mirrors real life Silicon Valley culture. Emily Chang's 2018 book *Brotopia* includes an interview with venture capitalist Chris Sacca, whose preferred mode of vetting new entrepreneurs was their ability to hang out in his hot tub, clearly a deal-making genre that disadvantages women. Chang writes: "Like those hot tub parties, much of the troubling behavior that marginalized or excludes women happens outside the office, including lavish, drug-fueled, sex-heavy parties hosted by some of

Silicon Valley's most powerful men, who cast the odds in their favor by inviting twice as many women" (p. 12). For Chang, women went from being icons, as in the case of Grace Hopper, to outsiders, "profiled out" of the antisocial, mathematically inclined male stereotype that has "proved far too ubiquitous to change" (p. 22) and became a fait accompli by the 1960s when the industry began to rely on aptitude tests to hire programmers based on male personality profiles.

Fast forward forty years, and this history accounts for why today, women are the elusive minority in the real and the simulated Silicon Valley, where the fictitious bros live together as they try to promote their respective mobile apps, including Nip Alert. As the name suggests, the app is designed to locate women with erect nipples, which mirrors the controversial app Titstare, which was introduced at the 2013 TechCrunch Disrupt Conference by two Australian programmers as a fictional app, meant to be satirical commentary but commentary that led to charges of misogyny on the part of the programmers, the conference organizers, and the industry as a whole. The home's owner, Erlich, who sports a t-shirt reading "I know HTML: How to Meet Ladies," becomes exasperated that Richard isn't able to develop a more innovative app and threatens to evict him, which spurs Richard to pitch his app to the fictional tech titan Peter Gregory. Pied Piper, originally designed to be a music app that ascertains copyright status on downloaded files, contains a powerful compression algorithm that maintains file quality, and Richard finds himself in a bidding war between Gregory and Gavin Belson, the former for a $200,000 investment and 5 percent ownership and another for 10 million dollars to buy him out. Tech titans are parodied mercilessly in the series, as the real heroes are the programmers and engineers, with Richard favoring Steve Wozniak over Steve Jobs, in that "Jobs was a poser who didn't write code; he only packaged it."

Richard ultimately decides to involve his incubator roommates in his start-up and to not sell his app, concluding that "For thousands of years, guys like us have gotten the shit kicked out of us. But now for the first time we are living in an era where we can be in charge and build empires. We are the Vikings of our day." While he favors collaboration, the passive and shy Richard is a product of his culture, locating power in a masculinist metaphor of conquest. Not unlike the MAC vs. PC ad campaign I discuss in chapter 1, women are again peripheral, figuratively and literally, and often victim to a range of misogynist stereotypes. This includes the image of an older woman riding her bike as Miss Palo Alto 2014 to the blond cupcake app developer at a TechCrunch competition clad in a pink hoodie who knows nothing about Java, gets one guy to program her site, and tries to sleep with another. The Pied Piper team almost loses the competition because the unlikely sexual magnet Erlich sleeps with both the former and now current wife of one of the judges. On a side note, art and life merge around the character of Erlich, portrayed by the actor-comedian TJ Miller, who left the show in 2017 as part of a mutual agreement between the actor and HBO (Sager, 2018). Before and since his departure, Miller has had his own share of controversy, from his comments to women at the 2015 TechCrunch Conference, including calling at least one a bitch (Tiku, 2015), along with his 2016 arrest for assaulting an Uber driver, his 2018 bomb threat made while on an Amtrak train, and allegations of past harassment and assault of a woman when he was an undergraduate at George Washington University.

Richard's ability to focus during the TechCrunch competition is limited because of his fixation on a former love interest, but he manages to recover and save the day and win the competition. Sex is an obsession but women are inevitably distractions, for the most part occasional adornments. In *"Silicon Valley* and Responsible Satire" Will Butler (2014) reports that at a cast press conference, a woman who identified herself as a former Silicon Valley worker, asked "Are there any plans to have female characters down the line who are engineers?" For Butler, and for those who have worked in the tech industry, the series is realistic to a fault, with Butler questioning the extent to which Silicon Valley, with some episodes directed by women, has "the responsibility to portray anything other than the familiar, nerdist, male-dominated tropes that still hold back Silicon Valley the place." Throughout its early seasons the show hasn't altered the formula that has brought it to its current season. There are several stronger female characters that emerge in the series, from the executive assistant Monica, who pivots between her status as an ambitious, professional businesswoman and potential love interest, the new CEO Laurie Bream, who not unlike her realtime counterparts, is depicted as somehow less competent and visionary that her former male counterpart, and even the character of Carla, hired in season 2 and simultaneously resistant to both misogyny and feminism: "I am not a woman engineer, I am an engineer." Such viewpoints mirror the hopes of women to be taken seriously in a brogrammer culture, yet as with Mayer's comments about growing the Silicon Valley pipeline without regard to gender balance, both the rhetoric of Silicon Valley the series and the realities of Silicon Valley the culture do nothing to change the gender-power dynamics of the industry, nor the racial dynamics.

Erica Joy's (2015) "The Other Side of the Diversity" presents a compelling narrative of the pervasiveness of Silicon Valley bro culture and its impact on her lifestyle choices and inevitably her emotional well-being, as she focuses on her experiences navigating various tech cultures, the "psychological effects of being a minority in a mostly homogenous workplace for an extended period of time." Joy's experiences as a black woman include numerous microagressions, from racist and sexist jokes to pervasive harassment by a male co-worker during her time as an IT field technician at Google. When Joy complained, rather than addressing the problem, Google transferred her. For Joy, companies like Google place great emphasis on improving their demographics without considering the behavioral changes that are necessary. Describing the impact on her physical and mental health, Joy chronicles how the constant pressure for workers to assimilate limits companies to more effectively integrate. Joy shares her attempts at better work-life balance to recover her authentic self and her attempts to engage in activist outreach initiatives, including Black Girls Code and Hack the Hood. Joy's is a powerful technofeminist storiography, a rewriting of the story of assimilation so many women, especially women of color, are actively resisting: "I am not my job. I am not my industry or its stereotypes. I am a black woman who happens to work in the tech industry. I don't need to change to fit within my industry. My industry needs to change to make everyone feel included and accepted."

Joy's storiography suggests that while the rhetoric of change is a dominant one within Silicon Valley, the reality of change continues to be a challenge, most recently evidenced through an email from former Snap (parent company of Snapchat) software engineer Shannon Lubetich to her engineering team on her last day, calling for a more caring, welcoming culture for women and people of color in response to what she has since revealed was a "pervading sexist vibe" that included "dick jokes" and the hiring, similar to Microsoft, of scantily clad models and dancers. Interviewed by Kara Swisher during the Code Conference 2018 as part of an episode of MSNBC's *Revolution: The Tech Titans Shaping the Future*, Snap's CEO Evan Spiegel acknowledged the email and its public aftermath as a wake-up call (Recode Staff, 2018), leading to changes in executive communication protocols in ways that honor diverse perspectives at a company whose female workforce is less than 25 percent at the management level and 13 percent at the technical level (Cheddar, 2018).

Of all the fictional TechCrunch presenters in the *Silicon Valley* episode I cite, only one group is female, and this may reflect the not so fictional statistical data of the lack of interest in women's startups. As Sissi Cao (2018) reports "While women own 38 percent of businesses in U.S., only two percent of venture capital went into women-led startups." Cao cites Jessica Walker, Manhattan Chamber of Commerce president, whose assessment is that women would be better supported by relying on crowd funding rather than the typical venture capital model common to Silicon Valley. Walker affirms that "I think women have an advantage in telling unique stories, either about their companies or about themselves, that help them in getting capital in crowdfunding." The problem is significant enough that numerous activist organizations have evolved, including Lesbians Who Tech and Tech Ready Women.

CONCLUSION: MAKING THE PROBLEM VISIBLE

Despite that initially exasperated read on the long plane ride out West, I reread *Lean In* through a slightly more positive lens thirteen months later, several months after concluding a highly stressful administrative role as a department chair and at a time when I benefitted from my own class privilege in the form of a paid administrative leave, an opportunity to lean back that many women, even in the academy, do not possess. And perhaps that is the inevitable problem with *Lean In* as a supposedly feminist text: Sandberg, perhaps rightly or wrongly, identified her primary audience to be women "fortunate enough to have choices," those privileged enough to decide when and if to "lean in" (p. 8). As a result, *Lean In* may represent an empowering narrative for some women, but it is a white, upper-class narrative that neither accounts nor apologizes for the diversity of experiences it excludes.

For that reason, it is vital not to presume that any discussion of women's histories or women's work in the field represents a singular experience. It is important to make visible the differing experiences of women in a spirit of equity and access, even acknowledging the power dynamics that empower some and alienate others often on the basis of not only gender but also race, class, sexuality, and ability as well. While both Abbate's (2012) and Sandberg's (2013) texts, which I reference early in this project, represent important opportunities to talk back by providing counter-narratives of the IT industry, we must understand that women's (his)stories of technology should move beyond the individual to the collective and from the local to the global. In this way, a technofeminist perspective becomes a form not of complicity with more dominant narratives but a form of resistance and political action. As part of his interview for the MSNBC *Recode* special, Sundar Pichai indicates the need to release women who have experienced harassment and other forms of workplace hostility from the non-disclosure agreements that keep them from sharing their stories. Pichai does not reference the current #MeToo movement dominating social media as women share and/or empathize with the trauma of misogyny and harassment across entertainment, political, and business arenas.

In late 2017, notable women in the entertainment industry were collectively named *Time Magazine*'s Person of the Year as The Silence Breakers (something I will discuss in chapter 4), an indication that even as individuals share stories of misogyny and assault, the power, as we have seen with the 2017 and 2018 women's marches around the globe, is in the collective. Both the rhetorical dramatizations and the increasing real-life narratives of women across tech industries suggest that #MeToo is equally applicable to Silicon Valley. Part of that narrative involves continued gaps in pay as well as representation, something I alluded to in chapter 1. And for all of Bill Gates's philanthropy, Microsoft has been touting data that prove pay parity between men and women but has failed to account for a male-dominated hierarchy in which men outnumber women by five to one (Day, 2018). There are efforts to address this inequity, as a *60 Minutes* profile (Hartman, 2018) on the enterprise cloud computing company Salesforce featured CEO Mark Benioff as determined to maintain efforts to balance pay inequities through salary increases, promotions, and overall inclusion at the table. The challenges continue to occur; Benioff concedes that as the company grows more successful and acquires other cloud-based customer management companies, they acquire their cultures as well, ones that continue to discriminate in both representation and salary. This phenomenon mandates continued corrective action, a process that can and should apply to other aspects of the IT industry but typically does not, as companies both large and small "talk the talk but do not walk the walk."

As my next chapter documents, there have been numerous attempts through artifacts marketed to children and adolescents to create more collaborative, gender-fair depictions that counteract the "complicated genius" leadership of a Steve Jobs or a Mark Zuckerberg and the male-dominated technology and entertainment industries that privilege that mythos that excludes women. The narrative of tech genius, white males who have an idea for a product, a network, or a service continues to reign. Molly Lambert's (2015) "The Difference Machine: Ada Lovelace, Grace Hopper, and Women in Tech" calls for a complication and reversal of this narrative to move from not just male to female but from the individual to the collaborative:

> one major problem with biographies about difficult men is that they tend to imply a connection between their subjects' importance and their self-importance, downplaying the structural advantages that helped put them in a position to claim so much credit." To combat the concept of the tech bro, there must be a tech sisterhood. Tech history is not a chain of command, it's a crazy quilt—no machine is ever really built by one person alone. It would be a mistake to consider Ada Lovelace and Grace Hopper as just lone geniuses—the same way it is a mistake to think that way of the men. But they are two avatars of importance for women in tech—the proof that natural talent knows no type. (Lambert, n.p.)

Ada, Grace, and Hedy had those ideas too, sans a tech sisterhood, in respective eras that made them innovators ahead of their time. If we made that legacy more visible, along with the legacies of their more diverse sisters of color, then perhaps that would serve as an educational catalyst. Certainly, their legacies are made more prominent in museums and in documentaries, as the website for the Computer History Museum features the exhibit "Thinking Big: Ada, Countess of Lovelace," jointly sponsored by the Association for Computer Machinery and Google; and Alexandra Dean's 88-minute documentary *Bombshell: The Hedy Lamarr Story*, co-produced by actor Susan

Sarandon. In his review of the documentary, Dennis Harvey (2018) contends that the film "skims over Lamarr's more troubling and troubled aspects to paint her in some stock terms as the victim of keep-her-on-that-pedestal misogyny," and calls for a more balanced portrait. Harvey ultimately concludes that "Many a highly intelligent person has led a messy life. Aiming for the inspirational, *Bombshell*'s revision tilt inadvertently ends up reducing its subject by exalting her." Such an assessment typically has not applied to the likes of Steve Jobs, whose mythos has been so culturally imbued that his dramatic flaws as a human being are frequently glossed to secure the legend, as we also see with the depictions of Alan Turing and Steven Hawking that introduce this book. Although Harvey's review may not be a representative assessment of the film itself, his assessment of Lamarr lacks awareness of his own misogyny as he questions the hypocrisy of wanting to be known for more than just a beautiful face but to then to become "a full-blown plastic surgery addict." Harvey determines that Lamarr's presumed resistance to and complicity with her status as a sex symbol is the result of her own troubled personality rather than a culture that celebrates feminine youth and beauty and discards intelligence and age, just as we see in the reaction of those not so fictional *Silicon Valley* bros to the image of an older white woman. Yet toward the end of her life, Lamarr was still inventing, including a mobility aid for older adults to get into and out of bathtubs, a fact she was proud of in recordings of her that were featured in the 2004 documentary *Calling Hedy Lamarr* that I cite in chapter 1.

In multiple episodes of *Silicon Valley*, a Steve Jobs bobblehead doll figures prominently, bedecked in the classic black mock turtleneck; if only there were a Hedy Lamarr bobblehead doll, in the same way there were and are vintage Hedy Lamarr paperdolls available on sites like eBay to adorn with gowns and shoes, perhaps her legacy would be more like the GE advertisement promoting the legacy of Millie Dresselhaus I feature in my previous chapter. And perhaps if Silicon Valley promoted collaborative models of technological innovation at Facebook and other companies that do exist among employees, including women employees, the pipeline for women would expand, and we would see more gender balance represented on the screen, on the page, and in real life as women became more visible in such collaborations. Emily Chang (2018) concludes that the "scarcity of women in an industry that is so forcefully reshaping our culture simply cannot be allowed to stand" (p. 14). Silicon Valley needs to be more inclusive, aligning with Marissa Mayer's call for the "technological future for our world that we really deserve" (qtd. in Chang, p. 14). Without such efforts, the vicious cycle of fact vs. fiction provides a limited future for women and girls and the future of the tech industry as well, as women need to see and benefit from their historical legacy of innovation so that their opportunity for leadership is equal to their male counterparts. As Hedy Lamarr aptly said when she learned she would receive the Electronic Frontier Foundation Award in 1997, three years before her death, "It's about time."

Chapter Four

Gender Play and the Marketing of Misogyny

Lois Gould's 1970 "X: A Fabulous Child's Story" satirizes society's obsession with and construction of gender through the tale of a child whose status as male or female is masked behind the label of "X." With no ability to gender names, clothing color, toys, and physical activities, the friends and family of Baby X's parents, who have agreed to the gender neutral "Secret Scientific Xperiment," grow increasingly frustrated at the thwarting of the identity performances that gender represents. By the time Baby X enrolls in school, the problem compounds in that other children ostracize the child because of its lack of conformity to gender roles for which they have been socialized to perform, even at such a young age. Despite these challenges, the story ends happily, with X finding acceptance and becoming a diverse role model to other children: boys who want to do needlepoint and vacuum the carpet and girls who want to play football and mow the lawn.

As someone who was born in 1963, I identify with Gould's story because of the gendered nature of play and the equally gendered mythologies that circulated through toys and games aimed at boys and girls. Later, as a college-level writing instructor in the 1980s and early 1990s, I often assigned the story as part of a unit analyzing gender codes inherent in toys and games, sending students on field research to the local Toys "R" Us to find relevant examples that documented assumptions about girls' play versus boys' play. Certainly, my own childhood experiences weren't much different: my parents purchased a toy stove, a model grocery store that folded up like a suitcase so I could learn to shop for plastic miniature steaks and boxes of peas, and the ever-canonical Barbie doll, though I was actually given the African American Christie as a first Mattel Barbie. And although I desperately wanted to play with Hotwheels and only was able to do so when we visited the parents of two little boys who lived down the street, my parents also purchased other gadgets and games: a tape recorder and an early favorite board game, Careers, produced by Parker Brothers and designed by sociologist James Cooke Brown (1921-2000) to presumably foster interest in employment paths in the Arts, Ecology, Politics, Big Business, Space, and Teaching. In the game, at least, I always wanted to be an astronaut. Despite the gendered nature of the playthings I received and by that point, desired—one of my all-time favorites was, sadly, the boardgame Mystery Date—my parents were encouragers of technology and they ensured I received a typewriter to help with my high school work and preparation for college. It also helped land me a job as a student secretary in the journalism department, where I was also an undergraduate major.

Similarly, my own involvement with computer technology is one that began in a very gendered way; in my student secretary role in the early 1980s I first used a computer at that time to do mail merge files for envelope labels. I found the VT100 I used very unintuitive, but I was fascinated at being able to merge the files and to ultimately send an email message across a campus network. When the journalism department installed a microcomputer lab of Sanyo PCs, each with a big 256K, I was asked if I wanted to help maintain the lab; it was either this or be a teaching assistant in physical anthropology. At this time in my life I seemed to love computers and bones, but I ultimately selected the computer opportunity, and this led to my own computer purchase, with the help of my mother, of a comparable Sanyo system to type my papers in literature, philosophy, and journalism. I remember so well the 5-and-a quarter-inch floppy disk and the bright green monitor; this system was soon replaced though when I became a graduate student with a larger PC, this time purchased by a boyfriend. In many ways, my literacy sponsorship has been influenced not just by gender conditions but by economic conditions as well.

Nevertheless, little did I know that my later time as a graduate student both at Sacramento State and at Purdue University would solidify my interests in teaching with computers, from the Mac IIEs I used in learning skills labs at Sac State to those tiny Mac Classics I used at Purdue University to teach technical communication courses in the early 1990s. My own research interests were more feminist, particularly involving the image of women in popular culture artifacts such as advertising, and as that imaging process migrated to the graphical user interface of the web, I found my research and teaching interests migrating along with them. I wanted to have the power to represent myself online, so I painstakingly learned .html from a male systems administrator in my first tenure-track job so that I could have students develop similar projects. I remember teaching .html to undergraduates in very mnemonic ways; B is for Bold, I is for Italics, but my skill sets enabled me to be put in charge of developing a range of websites for the division of arts and humanities, work that I initially loved, but work that I began to feel was almost secretarial in the way I was expected to update and fix files on a 24/7 basis. In fact, when I left that first job, some of the .html editing did in fact go to some of the secretarial staff, in a process of deskilling that represents a historical move as technologies become gendered, including as I've noted in chapter 2, programming. In my case, .html was a skill others didn't have time to learn. When I arrived at my next job, that skill set soon put me in the position of being the "webmaster" for the English department website, a role I played for a very long time. Because this book focuses on the power of women's stories to speak back to more dominant histories of computer technology, it seems natural for me to share my own. And based on my computer literacy history, and my academic affiliation with the field of computers and writing, I've become very attuned to women's relationships to technology and the role of narrative in articulating that lived experience, something that influences the work I've done inside the classroom with literacy narratives as well as in articulating the possibilities and constraints for women in digital spaces. This includes, as I note in this chapter, the ability of women to use literacy technologies to talk back to more dominant male narratives of heroism and genius in the larger culture. I will make further connections to computer coding and digital literacy curricula in my final chapter, notably in the efforts to make STEM education more equitable.

STILL TRYING TO "DISRUPT THE PINK AISLE"

Technology then and now continues to be gendered as male, but how can we make visible the ongoing experiences and contributions of women? Certainly, my own childhood and early adult experiences represent a snapshot in time, but such gendered assumptions span the generations. In games like Girl Talk and Girl Talk Date Line, Mall Madness, and Electronic Dream Phone, the themes always revolve around boys, shopping, gossiping, and, again, boys. Even my childhood favorite, Careers, morphed by 1990 into a girls' version, Careers for Girls, featuring the options of Super Mom, Rock Star, School Teacher, Fashion Designer, and Animal Doctor. A current site titled GirlGames.com, complete with pink interface, features the categories of Frozen (where youngsters can play Anna and Elsa games), Barbie, Dress Up, Cooking, Makeover, Love, and Baby. Pink is pervasive. In their article "Pink Gives Girls Permission," Erica Weisgram et al. (2014) contend that

> Gender typing of toys, explicit labels, and gender-typed colors are cues that children may use to classify stereotypes about which gender should play with them. . . . The toys with which children play can serve to create experiences and environments in which physical, social, and cognitive development occur

and thus marketing is not only about making money, but also about marketing environments and experiences to children. (p. 402)

The results of their study conclude that the color variable is significant and that children have been socialized to the extent that the color pink impacts whether a toy was seen as appropriate, with pre-school girls choosing the pink toy whether it was labeled for boys or for girls. Such visual, semiotic features have led to numerous calls for more gender-neutral toy development and marketing, most notably the British consumer advocate group "Let Toys Be Toys," which has put international pressure on companies that include the now bankrupt conglomerate Toys "R" Us to eliminate gender labeling on toy packaging and store aisle location. As Megan Perryman, a Let Toys Be Toys representative stated, "boys and girls are still growing up being told that certain toys are for them, while others are not. This is not only confusing but extremely limiting as it strongly shapes their ideas about who they are and who they can go on to become" (Delmar-Morgan, 2013). Since its inception in 2012, Let Toys Be Toys has won multiple international awards and has influenced and extended to other campaigns, including "Let Books Be Books."

Clearly, disrupting the pink aisle and the gendered assumptions about girls and women that go along with it, is a longstanding need. This chapter addresses that need by extending my introductory overview in chapter 1 of early childhood, adolescent, and young adult media and technology artifacts, that in some cases reinforce, but in others speak back to harmful stereotypes of gender and technology. In chronicling these assumptions, I will also stress that these gender dynamics as they circulate within early childhood and adolescent media become longitudinal, transitioning to adolescent and young adult activities such as gaming. From a technofeminist perspective, there are material and economic consequences for women and girls as those assumptions circulate from the gameplay to the information technology industry, both for the workforce itself and for the consumers who are drawn to it. Despite an increasing emphasis on positive images of women's and girls' relationship to technology, these counter-rhetorics are in ongoing competition with a reality that continues to devalue women's current and future contributions to the material history of technology innovation and its representation in the larger culture.

In her research on gender bias in STEM fields, Rachel D. Robnett (2016) reports the result of surveys of female high school, college, and doctoral students about the form and frequency of gender bias and its overall impact on their self-concept as STEM students and professionals. Very often the strongest contributing factor involves numbers: students are often socially isolated in their lack of a peer group, and Robnett's findings support earlier research about the extent to which women seeking agency in arenas where men have traditionally dominated encounter misogyny and backlash. Thus, it should not be surprising that Robnett's data support the view not only that the greatest frequency of bias occurs in math-intensive fields, such as computer science, but also that the most prevalent type of bias is through interpersonal contacts with male peers. In addition to calling for future longitudinal research to more generally gauge the level of gender bias in specific STEM disciplines, Robnett also advocates for interventions that mitigate these definitive findings in which 61 percent of participants reported gender bias in all three educational levels in her study. For Robnett, and many other feminists, this finding mandates more emphasis on creating a culture in which women and girls have a supportive peer network that includes more gender balance in both student, faculty, and administrative roles.

GENDER TROUBLE IN SILICON VALLEY

Although Robnett does not extend her discussion to the STEM workplace culture, she does cite research that notes that such environments can foster negative employee perceptions of their own performance, their actual performance, and supervisor assessment of those performances. These perceptions can lead to both material and cultural consequences for women working within the IT industry, including the ongoing perception that they are outsiders and that their intellectual contributions are literally of less value organizationally and economically. There are numerous examples to support this phenomenon. In 2017, tech industry titan Google has had its share of gender troubles, first with the increasing pressure from the US Labor Department to provide employee data in response to charges about company-wide wage discrimination in which women are paid less than male counterparts at all levels (Levin, 2017). Media coverage documents that current and former women employees at Google are considering a class-action lawsuit. And then there is the "diversity manifesto" shared with fellow employees by software engineer James Damore, who argued that biological differences between men and women, including women's tendency to "neuroticism" and increased stress and anxiety, "may explain why we don't see equal representation of women in tech and leadership." Damore's damning conclusions about women's obstacles to professional advancement not only blame women for their disadvantaged status but pathologize them, given that the operational definition of "neuroticism," according to the current Wikipedia entry (Wikipedia.org, July 2018) for the term states that "Individuals who score high on neuroticism are more likely than average to be moody and to experience such feelings as anxiety, worry, fear, anger, frustration, envy, jealousy, guilt, depressed mood, and loneliness. People who are neurotic respond worse to stressors and are more likely to interpret ordinary situations as threatening and minor frustrations as hopelessly difficult."

Because Damore took most of his supposedly scientific data from Wikipedia, it is easy to see how the lack of in-depth knowledge and research, not to mention his own gender bias, would lead to a biologically essentialist set of conclusions as to why there aren't more women working at Google and in the IT industry in general. Moreover, as the economic research of Claudia Goldin (2002) has suggested, "discrimination against women is motivated, in part, by the desire of men to protect their threatened occupational status." Goldin's pollution theory of discrimination, or the perception that "new female hires reduce the prestige of a previously all-male occupation" (p. 31) explains but does not excuse the reactions of male colleagues in a broad range of historical and contemporary work roles and career paths. Although Goldin's contexts include manufacturing and service professions such as firefighting and law enforcement, intimidation and harassment as ways to restrict the entry of women and create barriers to hiring them in specific roles are common to a range of professions, including IT. Despite his privileged status as a white male within the info tech "broculture," Damore was promptly fired by Google, though cynics question if that was more a response to increasing cultural, legislative, and legal accountability for its hiring and personnel management practices. Damore soon after filed his own legal complaint and has granted interviews to anti-feminist YouTube channel hosts.

One compelling response to Damore and his critique of Google's diversity initiatives is GoldieBlox founder Debbie Sterling, who while acknowledging some of her earlier cultural assumptions that inadvertently could reinforce gender stereotypes ("boys like building, girls like reading"), publicly challenges Damore for doing the same (Sterling, 2017):

> most of your manifesto reads this way and what makes it worse is that you are a very privileged white male who has no idea what it feels like to go to work at your engineering job every day worried that your colleagues, your boss, potential investors, partners, etc. might be thinking in the backs of their heads that you don't have what it takes because of your gender. The reason girls and women aren't going into engineering and technology isn't biological. It's because guys like you make it an unwelcome environment. (Sterling, n.p.)

I have discussed GoldieBlox in earlier chapters, and while there are limitations to the eponymous lead character of the series, this campaign can represent positive models for girls, notably through the GoldieBlox book series by Stacy McAnulty, a mechanical engineer turned author. McAnulty has partnered with Debbie Sterling and illustrator Lissy Marlin to develop titles such as *Goldie Blox ~~Ruins~~ Rules the School* and *Goldie Blox and the Three Dares*, in which the young engineer and inventor collaborates with both female and male characters, including the African American Ruby Rails, the computer coder who loves technology and fashion, the latter as "a way to express yourself" (McAnulty, *Rules*, p. 31). Set in the mythical Bloxtown, named after Goldie's grandmother, the series certainly shows that Goldie Blox and Ruby Rails are as talented as their male counterparts and as entitled to the labels that have typically accompanied technology innovators: "creative genius" and "computer genius" and more pejoratively "gearheads," all labels within the *Rules the School* book. But I would argue that this type of labeling, despite McAnulty and Sterling's attempts to make technology a gender-fair enterprise, risks the reinforcement, as I noted with the title character herself in chapter 1, of stereotypes, in this case of the technology innovator as lone genius, often the male computer geek.

Naturally, the tech industry is like any other workplace culture in which employees develop collegial, interpersonal relationships with co-workers, developing community and support networks on and off the clock. But because of those longstanding cultural assumptions about the demographics of that culture, in growing numbers, women are reporting their inability to secure these networks, sharing firsthand accounts of not fitting in despite having the technical and educational credentials to do so. As former software engineer Megan McArdle responded to Damore's memo, articulating how the "brotastic culture of IT" led her to exit the workforce not because of explicit sexism or wage discrimination but because of her perpetual status as a marginalized minority:

> As my story also suggests, when a field is mostly guys, it's going to feel less than perfectly comfortable for women unless some pretty heroic efforts are made to counteract all that free-floating testosterone. That may retard both women's career prospects and their interest in joining that field in the first place. So even if the disparities don't start off as discrimination, you can still end up with an environment in which women who could be great engineers decide they'd rather do something else. A "natural" split of, say, 65-35 could evolve into a much more lopsided environment that feels downright unfriendly to a lot of women. And the women who have stuck around anyway are apt to get very mad indeed when they hear something that seems to suggest they're not experiencing what they quite obviously are. (McCardle, 2017, n.p.)

McArdle identifies a "bro-ecology" that is longstanding, from the days of the Grace Hopper and the ENIAC programmers to Google, Apple, Oracle, and others. And as evidenced in the 2016 presidential election, a group of vocal young white male advocates for Bernie Sanders became known as "Berniebros" for their social media harassment of supporters of Hillary Clinton, or those who critiqued Sanders's record on racial injustice, and whose actions were comparable with the well-known sites 4Chan and Reddit. As such examples document, bro-ecology can escalate from an implicit but palpable culture of male privilege to one of violent misogynist rhetoric that threatens violence against women who attempt to infiltrate these cultures. Damore's manifesto is not one that explicitly threatens women's safety, yet it does represent exclusionary viewpoints and assumptions that threaten women's sense of security and emotional and economic well-being. Rather than accuse women of being essentially prone to "neuroticism," it creates a culture in which anxiety among women would be a likely result of such a system of male privilege, a boys' club that women are discouraged from entering. #gamergate, though not the focus of this discussion, is a notable example of that phenomenon, but it extends into other arenas, including the comic and superhero franchise culture, both for the women who work there and the cultural shifts on the big and small screen (and page) that place more women as the leading hero.

Consider the case of Heather Antos, a female editor at Marvel Comics, who in 2017 became the target of sexist harassment by Marvel fans for posting a Twitter selfie with six other Marvel employees drinking milkshakes. Not unlike the Google diversity controversy, similar misogynist and racist rhetoric has evolved in comic culture, with David Gabriel, Vice President of Sales for Marvel, claiming that diversity is to account for the decline in sales. *Telegraph* writer Adam White (2017b) summarizes that situation, stating that "While Gabriel was quick to backtrack on his comments, they remain a troubling example of the struggles facing storytelling for an audience seemingly willing to blindly accept superhuman powers and inter-dimensional portals, but not so much a diverse assembly of characters" who manifest those powers. Many of the "new" superheroes are in fact women of color, from African-American Riri Williams as Iron Man and the Pakistani-American Kamala Khan as Ms. Marvel, and White's contention that Marvel's declining sales are more likely due to the rising costs of comic books when crossover plotlines ensure the need for multiple purchases.

Regardless, women are rising up against the backlash; in the case of Heather Antos, groups of women began posting selfies with the hashtag #makeminemilkshake (I shall talk more about the power of such hashtag activism in my final chapter). Other powerful forms of talking back to traditional male storylines is through cosplay culture in which more and more women are taking on superhero personae. I focus on comics here because the culture of comics, as with the umbrella culture of IT, promotes a male adolescent stereotype complete with acne, braces, glasses and pocket protectors, as in the recurring character Comic Book Guy, the adult manchild from the longtime animated series *The Simpsons*. As if the cultural stereotype were a fait accompli, the comic book industry has been as male dominated as the video game industry and general information technology workforce. As with the depiction of technological and scientific innovation in the larger culture, figures like Steve Jobs become larger than life deities, the superheroes of the tech industry if you will, in ways that mirror the predominantly male superheroes of both the small page and the big screen and that reinscribe a masculinist culture that excludes women and makes their presence abnormal in both workforce contexts. When women are present, their use value has often been limited, with hypersexual depictions in comics and video games, notably in older versions of the Ms. Marvel and Lara Croft comic series (Hickson, 2016), and the ubiquitous strippers and prostitutes of the Grand Theft Auto series. As the hypersexualized Jessica Rabbit of the 1988 film *Who Framed Roger Rabbit* laments, "I'm not bad, I'm just drawn that way." Nevertheless, the emphasis on women as sexual objects extends to real life, given that Heather Antos and her coworkers were threatened with rape. Fortunately, there is evidence of a substantial cultural shift in recognizing the important role of women; at the 2018 Comic-Con International, novelist and comic writer Majorie Liu was the first women to win the Best Writer Eisner Award since its inception thirty years ago for her series *Monstress,* as well as Eisner Awards for best Continuing Series and Best Publication for Teens with *Monstress* illustrator Sana Takeda. Similarly, the team of Roxane Gay, Ta-Nehisi Coates, and Alitha E. Martinez, who won the Best Limited Series award for their collaboration on *Black Panther: World of Wakanda* (Ducharme, 2018). I shall further discuss the relationship among gender, race, and technology in the *Black Panther* series in my final chapter.

TECHNOFEMINIST REMIX PEDAGOGIES

In my own curriculum and pedagogies, I have encouraged students to develop and apply critical cultural frameworks that emphasize identity politics—and their production, distribution, and consumption—to the rhetorical analysis of historical and contemporary media artifacts and genres, including social networking, games, and other forms of digital media, as well as more traditional mass media such as advertising and television. A core reading for such efforts includes Lawrence Lessig's (2008) *Remix* and its emphasis on the evolution, primarily through social media and the emphasis on sampling, of a read-write (RW) culture in which citizens produce as much of their contemporary culture as they consume. Thus, remix becomes a powerful concept through which to talk back to dominant hegemonic practices via genres and modalities that foster such social commentary within a range of citizen vernaculars.

To encourage the socially conscious role such remixes can play, I have asked students in graduate seminars on media and cultural studies courses to complete the following assignment:

The Remix Project

Context:

We've seen them all around: images, videos, films, and other genres that sample an existing artifact and parody it for humorous or critical social commentary. This project asks you do to do a similar type of remix not only to experiment with mixed media production and composing tools but also to consider the ways such critical cultural work might play out in your own classrooms. So in other words, you can't just remix for remix sake, but rather select an artifact that resonates with you in some way and "talk back" in counter-narrative form to its larger cultural assumptions (perhaps about race, gender, class, sexuality, ability, and so forth) and to the individuals, groups, and corporations that disseminate those assumptions.

Genre:

You might remix a logo, an advertisement or commercial, a board game, a magazine cover, a map, a movie trailer or poster, an NPR essay in audio or visual form, a Wikipedia entry, even an Instagram selfie!

Resources and Tools:

The project can be as technologically low-end or high-end as you want, meaning you do not necessarily have to rely on digital tools to complete the task. This is dependent on what you want to remix and why and how to replicate aspects of the medium in which the original was produced. Our course space has a section devoted to remix genres and tools, including advertising, games, art and photography, magazines, and film/video.

Assignment Components:

In addition to your remix artifact, you must provide a brief statement that indicates what you're remixing and why. This written statement (app. 250 words, or equivalent) will accompany your submission of the assignment and serve as a brief presentation when we showcase them in class.

In one response, Lauren, a former high school counselor, chose to rely on a comic strip generator to create a series of panels devoted to the depiction of popular superheroes, including Ms. Marvel and Batwoman, by women of diverse body types engaging in cosplay. As Lauren reflects:

> More and more young women are reading superhero comics nowadays. In response, major publishers like DC and Marvel have begun creating new female characters and re-vamping old ones in order to meet demand for more female representation ...Many of these new stories also feature the intersectionality of their protagonists—Ms. Marvel is a teenager, a Muslim, and the child of Pakistani-American immigrants. Batwoman identifies as lesbian, and her relationship with Gotham City Police officer Renee Montoya is featured in her current arc. I wanted to highlight these characters in my remix project because even though these publishers have responded to the demand for diverse female characters, these representations are still far from the norm. The vast majority of superhero comics are still dominated by white, cisgender,

heterosexual men. Another issue I considered is that although the new crop of female superheroes brings more diversity to the genre, they tend to be drawn in overtly sexual ways with unrealistic, Barbie-like proportions. This realization led me to use photographs of female cosplayers. (O'Connor, 2015)

Lauren's project manifests the power of remix pedagogy to interrogate cultural constructions of gender and sexuality and to use digital media such as Photoshop and comic strip templates to perform a counter-narrative within common genres. Although this response to remixing gender and other aspects of identity politics occurred within a graduate seminar, I believe, as did many of the students, that the emphasis on remixing such cultural representations can and should occur both inside and outside the academy to foster that sense of technofeminist storiography. These remixes serve as counter narrative that helps to promote a sustainable ecology for women across the spectrum of the IT industry, including comics and gaming, whose male dominated culture has impacted the rhetorical representation of women in historical and contemporary artifacts and, as the Antos milkshake selfie documents, the reality of women's working lives in these digital industries.

Moreover, as Virginia Kuhn (2017) contends in her article "Remix in the Age of Trump," "For me, remix has always been a political issue: I see this work as subversive, as a way of speaking truth to power, a championing of the underdog, an occasion for thoughtful people to challenge narrow and damaging representations that comprise the visual hegemony of the Hollywood machine. Indeed, remix is a matter of free speech" (pp. 87-88). Ultimately, Lauren's remix is a retelling of a story of women and comics that aligns with technofeminist efforts such as Jessica Ivins, whose 2004 interactive site Retrotype allows women users, through keyboard commands, to recreate themselves as the video gaming industry would have represented women throughout what she has identified as five eras of gaming. As Ivin's artist statement explains:

> Retrotype's purpose is to chart, recreate, and question gender representation throughout the history of video gaming. Created for female users, it serves as both an interactive game and educational tool. The user creates a self-portrait comprised of features and components that resemble graphical representations of women throughout retro gaming. The user enters a self-determined year of her prime beauty, and is then presented with an interface displaying era-appropriate features for the creation of her digital self. When creating their prototypes, women must utilize key commands instead of clicking with the mouse. The deliberately forces women to participate in a less user-friendly form of interaction. It also mimics the actions of computer programming, which places the female in a powerful position as creator. (Ivins, 2004, n.p.)

THE SHARED SEARCH FOR OUR INNER WONDER WOMAN

The sexualization of women characters in comics and games has undoubtedly impacted women's and girls' ability to identify with such characters. Even the superhero Wonder Woman, who has finally made it to the big screen after seventy-six years, sans the tiara version of the character, is not without recent controversy. In 2016 the United Nations decided to brand the superhero an Honorary Ambassador for the Empowerment of Women and Girls, a short-lived title given the backlash from United Nations staff that resulted in a petition to "reconsider." As the petition's preamble reads:

> Wonder Woman was created 75 years ago. Although the original creators may have intended Wonder Woman to represent a strong and independent "warrior" woman with a feminist message, the reality is that the character's current iteration is that of a large breasted, white woman of impossible proportions, scantily clad in a shimmery, thigh-baring body suit with an American flag motif and knee high boots—the epitome of a "pin-up" girl. This is the character that the United Nations has decided to represent a globally important issue—that of gender equality and empowerment of women and girls. (Care2Petitions, 2016, n.p.)

An infographic (Mattick, 2016) published by Halloweencostumes.com documenting the various incarnations of the superhero's sartorial choices, with increasing amounts of both thigh and cleavage, visually supports this view. Yet, with the release of the 2017 film *Wonder Woman*, the backlash is one against the idea that a superhero film could feature a woman, even after three-quarters of a century, and be directed by a woman, Patty Jenkins. As Anita Sarkeesian notes (Petit & Sarkeesian, 2017) in her blog Feminist Frequency "I felt something stir in me that has always been there but rarely comes to life; the yearning, so rarely fulfilled, for images of women being the larger-than-life heroes men so often get to be, rather than just the ones who need the aid of a hero."

For many—but certainly not all women in 2016—that hero was Hillary Clinton. Perhaps it it is fitting that on October 27, 2017, her seventieth birthday, Clinton received the Women's Media Center Wonder Woman Award. For Clinton, Wonder Woman's big screen debut was long overdue:

> And honestly, it really is exciting to receive the first Wonder Woman Award. Yes! I, um, I saw the movie. I loved the outfit. My granddaughter was really keen on Wonder Woman, so I thought maybe I could borrow something from her for the night. It didn't quite work for me, but I will say that this award means a lot to me because as a little girl, and then as a young woman, and then as a slightly older woman, I always wondered when Wonder Woman would have her time, and now that has happened (Lindsay, 2017, n.p.).

Clinton's and Sarkeesian's stories resonate, for Sarkeesian because of the threats of violence she has received for speaking out in online forums about women in the gaming industry but also because of the ways our female tech superheroes have, until so recently, been relegated to supporting roles, as I note in my earlier discussion of Joan Clarke in *The Imitation Game*. In her 2017 book *Programmed Inequality: How Britain Discarded Women Technologists and Lost Its Edge in Computing*, Marie Hicks acknowledges that despite the cryptographer's accomplishments, she is "best known . . . for being Alan Turing's "beard" for a short while" (p. 37). Hicks's historical overview documents the ways that women's technical roles and responsibilities were increasingly deskilled post World War II, equating their labor to that of rote factory work until the rise of a managerial computing class of workers able, as Hicks notes, to go from machine room to boardroom, thus closing the door to many women. Thus, for Hicks, "all history of computing is gendered history" (p. 234), one that has continued to impact the dramatization of women's contributions to the rise of computer culture.

Similar to the concerns of other actors who find so few opportunities for empowering roles on the big screen, Reese Witherspoon's acceptance speech at the 2015 *Glamour Magazine* Women of the Year Awards outlines the typical supporting space for women in film:

> I want everybody to close their eyes and think of a dirty word, like a really dirty word. Now open your eyes. Was any of your words ambition? I didn't think so. See, I just kind of started wondering lately why female ambition is a trait that people are so afraid of. Why do people have prejudiced opinions about women who accomplish things? (Moeslein, 2015, n.p.)

Speaking of powerful women's roles, such as Goldie Hawn's *Private Benjamin* and Sally Field's *Norma Rae*, Witherspoon expresses frustration about the future for her own teenage daughter:

> She'd be forced to watch a chorus of talented, accomplished women Saran wrapped into tight leather pants, tottering along on cute, but completely impractical, shoes turn to a male lead and ask breathlessly, "What do we do now?!" I dread reading scripts that have no women involved in their creation because inevitably I get to that part where the girl turns to the guy, and she says, "What do we do now?!" (Moeslein, 2015, n.p.)

When male majority superiority is threatened, some men, as we see with James Damore, as well as the misogynist backlash against Hillary Clinton's presidential campaign, take it upon themselves to protest as women attempt to gain entry to male-dominated enclaves or to be empowered through their solidarity with other women. A notable example of such protest online and off occurred with the multiple all-women screenings of *Wonder Woman* at the Alamo Drafthouse movie theatre in Austin, Texas (CNBC, 2017).

CREATING A COUNTER NARRATIVE

If there is any silver lining to James Damore's now public statements on women's supposed biological unsuitability for IT careers, it is in the similarly vociferous amount of talk-back through the stories of women working in IT, confirming that it has been more culture than biology that has made their work in the industry challenging and in some cases untenable. Equally notable have been the public reminders about women's contribution to computer programming in historical, military, and workplace roles. In response to James Damore's mansplaining of the problem with diversity initiatives in IT and women's inability to compete, Holly Brockwell (2017) concludes that, "Damore clearly thinks he's schooling the world on biology, but it's actually history he should have been paying attention to. Because he either doesn't know or has chosen to forget that women were the originators of programming, and dominated the software field until men rode in and claimed all the glory." As part of her critique, Brockwell summarizes Ada Lovelace's, Grace Hopper's, and NASA's Katherine Johnson's and Margaret Hamilton's respective achievements. For Brockwell, "sexism is just bad programming," steeped in power and culture and not biology. Yet sexist assumptions such as Damore's continue to erase such female contributions from the history of computing, leaving this programming to the genius model of technological and entrepreneurial innovation.

Just as it is important to not essentialize women as innately predisposed to experience mathematical and technological anxiety, it is equally important to not essentialize men as innately predisposed to deny women's contributions or to engage in systematic gender discrimination across cultures and professions. Fortunately, women have not been the only ones to talk back. To further align this discussion of comic culture and technology innovation, I now turn to a series of comic and graphic novels that promote positive images of women, girls, and technology and acknowledge women's historical contributions to computer science.

I first begin with Gene Luen Yang. Yang is internationally known in comic and graphic novel circles for his book *American Born Chinese*, a National Book Award Finalist for Young People's Literature and recipient of other prestigious awards. As a child of Asian immigrant parents growing up in California, Yang ultimately pursued a degree in computer science, with a minor in Creative Writing, at the University of California, Berkeley. Yang's childhood dream was to be a Disney animator. *American Born Chinese* tells the story of Jin Yang, a Chinese boy who morphs into Danny, a white boy who assimilates into a white culture. Yang's compelling depiction of the stereotypes of Asians and the struggle for self-acceptance for adolescents and young adults has brought him great acclaim, including a MacArthur Fellowship or "Genius Award," one of only three graphic novelists to receive the recognition. Not unlike Ada Lovelace two centuries before him, Yang's work aligns the mathematical and humanistic, a combination that for Yang resulted in his early career as a computer science teacher in Oakland, California, for nearly two decades and a career as a comic artist and graphic novelist.

What also aligns Yang with the goals of *Technofeminist Storiographies* is his work with illustrator Mike Holmes on the now five-part series *The Secret Coders*. The series chronicles the story of the appropriately named "Hopper," an awkward, gangly, basketball playing redhead who moves to the new school with her mother, an Asian-American single parent who also teaches at the school. Appropriately, Yang's first novel in the series includes an author's note that serves as a computer literacy narrative, an account of how he became hooked on coding because of a summer school enrichment course in computer science, one of four courses his mother forced him to enroll in rather than stay at home in front of the television. Yang writes of being awestruck by the skills of another young coder named Bill, who made computer drawings via code in magical ways. Although it would be easy to lump Yang into the longstanding historical narrative of the genius model of technological innovation in his labeling of Bill as a "magician," Yang stresses the collaborative nature of the coding process, Bill's role as an early mentor who served as his class partner, and his family's decision to purchase a computer. Yang concludes his narrative: "Coding is creative and powerful. It's how words turn into image and action. It truly is magic. Mike Holmes and I made the book you now hold in your hands because we want to share a bit of that magic with you, and maybe inspire you to become a magician—a coder—yourself" (2015, p. 91).

What becomes clear from reading the *Secret Coders* series is that just as with Ada Lovelace and Charles Babbage and just as with Hedy Lamarr and George Antheil, technological innovation can be both collaborative and gender-equitable, as Hopper partners first with Emi, a boy of color who has three older basketball-playing sisters, and Josh, an initial bully who soon becomes a friend and partner in coding and decoding the mysterious history of the current Stately Academy where they are enrolled. Hopper, not unlike her namesake, is an outsider in a number of ways, but what unifies the three is uncovering the mystery of Stately's past as the Bee School, a technology academy that attempted to align humanities and technology through the creation of and interaction with robots, coded to be "instruments of art." Central to this utopic vision of technology is eponymous Mr. Bee, an Einstein-like character who prior to his current role as the Stately Academy janitor, was a professor, mentor and self-described "visionary." Consistent with the comic, super-hero genre, Yang's series follows a traditional story arc that juxtaposes good and evil, with a corrupt principal, adolescent rugby players as his henchmen, and a former student of Mr. Bee's, Pascal Pasqual, as the arch villain Dr. One Zero. But rather than reinscribe the idea that computer science and coding are for boys only, and geeky boys at that, Yang's narrative stresses that coding can be used to do good, programming robots to thwart attempts by Pasqual and Principal Dean to gain control of Professor Bee's menagerie of small and large machines.

Yang's series is meant to be educational as well. In his role as the Library of Congress's National Ambassador for Young People's Literature, Yang acknowledges his goal to align reading and computer-science education in the form of the interactive lessons peppered throughout the *Secret Coders* series designed to help students both encode and decode the processes that drive the robots' movements. Just as the evolving partnership of Hopper, Eni, and Josh mirrors the real-world collaborators of Grace Hopper and the teams of both the Mark I and ENIAC, the model creative genius, in this case genius gone awry, is aligned with Dr. One Zero/Pascal Pasqual, who is referred to as a "green-skinned coding genius" in the website's online promo for the third book in the series, *Secrets and Sequences*. The outcome of the series is an evolving one in which the coders will have to decode the access to the secret tunnels that house the "most powerful turtle in the world" to thwart One Zero, who has replaced Principal Dean as the school's leader.

My rationale for foregrounding Yang's series is its evident homage to Grace Hopper, its selection of a female protagonist, and its emphasis on a collaborative model in the act of programming. Equally important is the way in which Yang's goals are consistent with the educational vision of Hopper featured in chapter 2, and by extension the alignment of creativity and technology that comprised Ada Lovelace's own Byronic patrilineage. Yang weaves these threads in his characterization of Hopper, whose father Albert Gracie is one of Professor Bee's original students and who has disappeared from Hopper's life under equally mysterious circumstances. The connection between the creative and the technological is a notable theme, something that also powerfully manifests itself in the website that accompanies the series, including videos developed by Yang to teach kids to code using the computer language Logo. In addition, the *Secret Coders* website also includes a "turtle art" gallery in which students have used Logo or other languages to create often colorfully themed digital line drawings. A review of contest winners reflects a strong balance between male and female students, suggesting the ways that Yang's narrative has the potential to subvert the gendered hierarchies that keep women and girls from realizing their potential in the classroom and inevitably the IT workforce.

Equally significant is the African American character Eni, himself in a form of gender reversal as his three sisters are school basketball stars while he prioritizes his work with Hopper and Josh at the expense of his own athletic success, to the chagrin of his coach and his parents. In the fourth book in the Secret Coder series, *Robots and Repeats*, Eni faces increased pressure to give up his relationship with first Hopper and then both Hopper and Josh as his sisters portray Hopper as a bad influence that is risking Eni's chances to play basketball. Eni's sisters, in what they think is an act of protecting their brother's future as a sports hero, collude with Dr. One Zero and their parents to split the trio apart when their coding leads to discovery of a hidden entrance to the new principal's conference room containing his secret plans. Only Hopper's mother comes to their defense, arguing for the positive influences their collaboration and friendship has achieved. Ultimately, the relationship among Hopper, Eni, and Josh resists traditional gender assumptions, including the support both Josh and Eni display when Hopper is herself cut from the basketball team, with Eni chastising his sisters for their blunt assessment of the situation: "you got cut, girl."

These various role reversals make the *Secret Coders*, whether Yang would identify it as such, a form of technofeminist storiography that talks back to the dominant story of male genius, and a counter narrative that allows adolescents the opportunity to question the historical and contemporary depiction of computer science. That this narrative occurs in the form of a graphic novel is a reminder of the power of gender performance in all aspects of contemporary culture, as Yang's work is a notable response to to the male-dominated comic culture and a disruption of gendered assumptions about the appropriate activities for boys and girls. Lois Gould would have approved, as would have Grace Hopper, the ENIAC programmers Jean Jennings Bartik, Kathleen McNulty Mauchly Antonelli, Frances Bilas Spence, Frances Synder Holbeton, Ruth Lichterman Teitelbaum, Marlyn Wescoff Meltzer, not to mention Ada Lovelace before them.

Fortunately, for today's early childhood and adolescent readers, there are attempts to recover these now historical figures. The 2017 children's book *Grace Hopper: Queen of Computer Code*, written by Laurie Wallmark and illustrated by Katy Wu, tells Hopper's story as a role model for readers. They begin with her childhood curiosity for how various household gadgets work to her determined focus on her studies to secure entry into college. True to her life story, Wallmark and Wu chronicle Hopper's desire to contribute to the war effort, her work on the earliest computers of the twentieth century, including her discovery and coinage of the term computer "bug," her status as a team leader (thought admittedly less as a collaborator), and her return to service after forced retirement. Stressing her status as an unconventional, rule-breaking innovator, it is unclear why despite so many empowering labels within the book, the title reinscribes the emphasis on queens and princesses, except perhaps that it represents a gender role with which young girls can identify. While that regal status frames the title and the book's final passage, there are other qualities that make Grace "amazing." Featuring quotes from Hopper throughout the book, a timeline of her life and numerous awards, the book ultimately introduces readers to Hopper's many roles in her lifestory:

> Software tester. Workplace jester.
> Order seeker. Well-known speaker.
> Gremlin finder. Software minder.
> Clever thinking. Lifelong tinker.
> Cherished mentor. Ace inventor.
> Avid reader. Naval leader.
> Rule breaker.
> Chance taker.
> Troublemaker.
> AMAZING GRACE. (Wallmark, 2017, Front Matter)

Author Laurie Wallmark has degrees in biochemistry, information systems, and writing for children and young adults, while illustrator Katy Wu's credentials include Google, Pixar, and work in gaming and social media. I mention their backgrounds because in the spirit of technofeminist storiography, Wallmark dedicates the book to her two daughters, while Wu dedicates the book to her mother "and to the women who strive to make the world a better place for young girls everywhere." Wallmark's first book (2015) was appropriately about Ada Lovelace: *Ada Byron Lovelace and the Thinking Machine*.

Another notable example in homage to Lovelace is the Ada Lace adventure series, published by Simon and Schuster, where "eight-year-old Ada Lace uses her love of sciences and gadgets to help solve problems with her best friend Nina." Developed by a former NASA Engineer, Emily Calandrelli, who has degrees from MIT and is the host of *Xploration Outer Space*, the Ada Lace series grapples with the presumed tension between science and art through young Ada's status as an inventor of robots who happens to be colorblind and her father's status as her school art teacher, whom she feels she is not able to please in the book *Ada Sees Red* (2017). As I stressed in chapter 1, the science vs. humanities tension was present in Lovelace's life as she herself grappled with the legacy of her Byronic lineage and her mother's attempts to quash those passions in her young daughter, passions that reasserted themselves as Lovelace aged. Calandrelli herself bridges that gap, and as numerous examples suggest, women in STEM are speaking out across genres and modalities, with children's books being an important venue for young girls to see alternative options to the pink princess aisle of the local toy and bookstore.

An equally positive model also occurs in the Simon and Schuster's Ready to Read Series. Laurie Calkhovan and Alyssa Petersen's (illustrator) *You Should Meet: Women Who Launched the Computer Age* (2016), featuring the women of ENIAC. The 48-page children's book begins in the present day, asking readers where we would be without the computer devices, from the laptop, tablet, to wristwatch, that govern our lives, and quickly moves to the historical and cultural contexts that led to Jennings, McNulty, Bilas, Lichterman, and Wescoff's role as "human computers." The book provides brief biographies of the women, and makes clear both the difficulty of learning to program a computer and the lack of credit they ultimately received for it, including the example of the all-male celebrations I featured in chapter 2. In addition to providing a broader history of computing in general and women in computing in particular, *Women Who Launched the Computer Age* is undoubtedly a technofeminist text, educating young readers to the exclusion and absence of these important historical figures. Recognized with a Best STEM Book Award by the Children's Book Council in 2017, the book also includes the stories of ENIAC Programmer Project founder Kathy Kleiman and her work in recovering these women's stories: "She saw pictures of ENIAC with men and women standing in front of it. The captions for the pictures named only the men. Kathy asked who the women were and was told they were models hired to show off the computer. That didn't sound right to her" (p. 37). The book's final pages honor women's roles by including brief biographies of Ada Lovelace and Hedy Lamarr, a history of programming that honors Lovelace's role in writing the algorithm for Babbage's Analytical Engine and, like the *Secret Coders* Series, activities for children.

ONE STEP FORWARD, TWO STEPS BACK?

In its stated role of putting more than 20,000 women in technology-based jobs by the year 2020, General Electric has followed up on the Millie Dresselhaus commercial that I profiled in chapter 2 with a recent television commercial featuring the fictitious Molly, whom *AdWeek* terms an "inventive star." Viewers first see Molly as a bespectacled child, frustrated by her father's request that she take the trash out in the rain while she'd much prefer to focus on her inventive designs. These include a mechanical system using the backyard clothesline to move the trashbag to the can, evolving to other inventions that include an automated bedmaker, and a mechanical arm to turn the pages of her schoolbook. Viewers finally see Molly as a young woman who programs the robots to perform inspections at GE, both confident and proud when her African American male supervisor concludes "That's Amazing, Molly." Her response is a simple one: "Thank you." This is quite the evolution from the Microsoft Tablet commercial and original "I'm a MAC, I'm a PC campaign" I discussed in chapter 1, and indeed, Microsoft has developed the Make What's Next campaign honoring International Women's Day and that features women inventors that include Ada Lovelace. The website for the campaign opens with a video screen shot encouraging girls to "Change the World. Stay in STEM."

What makes these examples "heartening," as Angela Natividad (2017) depicts these women and STEM campaigns, is that via more global media genres and modalities such as websites, YouTube videos, and advertising campaigns, the stories of Ada Lovelace, Hedy Lamarr, Grace Hopper, Katherine Johnson, and others are becoming more visible across these mass distributed venues in ways that make their contributions as notable as Alan Turing, Stephen Hawking, and Steve Jobs. Natividad writes

> These efforts . . . not only focus on normalizing the notion of women in tech but on encouraging those same women to drop their muzzles of modesty. Because can you even name five female inventors or scientists? Apart from your Ada Lovelaces and Marie Curies, it's a hard sell. And that isn't to say we've been sitting on our hands (or using them to knit); women are responsible for many of the things we take for granted today, including liquid paper, windshield wipers, solar homes and computers. (Natividad, n.p.)

As these positive examples become more common within the larger culture, it is important to remember that part of the storytelling process, as my former student Lauren's comic book remix powerfully documents, is to equip women with the digital tools to tell these stories across media genres, including the graphic novels, advertising campaigns, and other media I've featured in this chapter. Although it is vital to engage in a revisionist "her"-storiography of women and technology, today's story must also be told by women themselves in ways that have the potential to subvert the dominant narratives of technological innovation. Yet the depiction of women in computing narratives, within the history of technology itself, to their visualized role in toys, games, comics, television, and film, reflects both cultural complicity and cultural resistance. Unlike the utopic Baby X, there is no "neither boy, nor girl" in terms of the assumptions of play, labor, and role. X and its sibling Y have a neither/nor option, a neutrality that rarely exists but must be part of a respect for transgender identities in the larger culture. Examples like the girlie Goldie Blox and the fashion-conscious Ruby Rails, or even Yang's Eni and Hopper, represent a both/and construction in which those who identify with a female gender identity can be techie and girlie, and both boys and girls can be techie and athletic.

CONCLUSION:
CONFRONTING THE PRINCESS PROBLEM

Is it acceptable to be both a techie and a princess? Our culture continues to struggle with a place for women and girls in the larger culture in general and in IT culture in particular. The Sony Pictures' *Emoji Movie* of 2017 documents this dilemma on the big screen. As the protagonist, Meh emoji Gene, sets off on a journey to escape the limits of his essentially blah role in the app world of textopolis, he and the formerly popular emoji High Five, seek out the character Jailbreak to lead them through the wallpaper maze surrounding the apps of the 14-year-old boy Alex's smartphone. Jailbreak, a tough female complete with blue hair and androgynous beanie, is making her way to freedom in the cloud of Dropbox, chased by bots set out by the female villain emoji, Smiler. Yet, the cool Jailbreak is soon revealed to be the lost Emoji Princess, complete with tiara, possessing not just the tech skills but the princess ones that allow her to help Gene save the day by disabling the bots. Gene develops a multifaceted identity beyond his essentialized "meh" response, one that brings together Alex and his love interest, and the movie concludes with Princess emoji reassuming her essentialized royal identity in a happily ever after textopolis, dreams of life in the cloud forgotten, and expected gender norms intact.

It goes without saying that such gender norms have consequences that limit the ability of women and girls to advance materially, culturally, and economically. And even as we take several steps forward, as many of the examples I've included document, there are continued misogynist obstacles along the way. As I conclude a first version of this chapter in later October 2017, news coverage is focused on the case in Massachusetts of a high school golfer Emily Nash (Domonske, 2017), who although she was allowed to play on the all-boys Central Massachusetts Division 3 Boys' Golf Tournament, was denied the winning trophy for her performance because she was female. The first place trophy was instead awarded to a boy who was four strokes behind her and the rules of the Massachusetts Interscholastic Athletic Association state that although girls can be part of a team, they cannot be recognized as individuals. Although this example focuses on athletic accomplishment, it is a story that connects to the goals of *Technofeminist Storiographies*: the need to let women's achievements as individuals and members of collaborative teams be heard, not hidden.

Some companies are getting the memo. LEGO, for instance, has released its Women of NASA set featuring astronomer Nancy Grace Roman, the now ninety-two-year-old former NASA executive scientist credited as the "Mother of Hubble" for her role in the development of the Hubble Space Telescope; computer scientist and the Presidential Medal of Freedom recipient Margaret Hamilton; the late astronaut, physicist and professor Sally Ride (1951-2012); and astronaut, physician engineer, and professor Mae Jemison, the first African American woman to go into orbit in 1992. The series is based on Maia Weinstock's proposal, which originally included NASA's Katherine Johnson, who did not grant approval. This is likely due to her partnership with Mattel (Watch The Yard, 2018) to include her likeness as part of the new Barbie "Inspiring Women" series, which also features aviator Amelia Earhart and artist Frida Kahlo. Senior Vice President and General Manager for Barbie stated that

> We know that you can't be what you can't see. Girls have always been able to play out different roles and careers with Barbie and we are thrilled to shine a light on real life role models to remind them that they can be anything. (Watch The Yard, n.p.)

Although my own 1960s Christie doll and the likeness of NASA's Katherine Johnson represent Mattel's attempt to market an alternative to a broader consumer audience, Alexis McGee (2018) contends that this is far less progressive than we think; in her discussion of Barbie's Dolls of the World Collection, which includes dolls from Brazil and Chile, McGee argues that the collection

> shows stereotypes in a more pronounced manner because Mattel taxonomizes and markets these dolls by country. Each doll is produced as and meant to represent a form of female empowerment from a different country, but underneath the pricy clothes and accessories "not every" consumer can afford, Mattel is producing replicas of stereotypes communicating an ideology that privileges consumption and American ideals. The company creates portrayals of the ideal woman based on a composite of female traits of by using dominant white American characters to create a number of dolls of other ethnicities. (McGee 2018, n.p.)

Certainly, in the case of Earhart and Johnson, their contributions to aeronautics are a representative of an American progress and conquest of new frontiers narrative that has long dominated our conquering new frontiers historiography. But would Mattel have selected either Earhart or Johnson had they not conformed to a feminine ideal?

Meanwhile, Weinstock's LEGO proposal ultimately garnered the necessary 10,000 supporters to move forward to a LEGO idea board and eventually being available online and in stores. Despite the fact that an Ada Lovelace and Charles Babbage idea submission also received the requisite 10,000 votes, the resulting review board determined that LEGO would not pursue the set, which included both Lovelace and Babbage figures along with the Analytical Engine and a proper Victorian tea party set. The idea was submitted by Stewart Lamb Cromar to honor both the two-hundredth anniversary of Lovelace's birth in 2015 and his father, a mechanical engineer, and was ultimately featured on the National Museum of Scotland's blog with a guest post by Cromer (figure 4.1):

Gender Play and the Marketing of Misogyny 97

Figure 4.1. Proposed LEGO Design for Ada Lovelace, Charles Babbage, and the Analytical Engine. Copyright © Stewart Lamb Cromar, 2016.

The LEGO submission not only would have added more gender diversity to the predominantly male minifigure series but also would have made the story of Lovelace's contributions visible to young people and LEGO fans worldwide.

When I began *Technofeminist Storiographies*, I was prepared to argue that larger cultural rhetorics needed to catch up with reality of women's contributions. Through the research and writing process, as well the recent increase in the number of toys, books, films, and advertising campaigns, we need more storiographies to counteract the perspectives of the many James Damore archetypes of technological broculture. To further combat this culture, I conclude this book with a chapter on the possibilities and constraints of various educational initiatives aimed at increasing the technological literacy of girls and fostering a sustainable ecological future for women and girls in computer culture.

Chapter Five

Sustaining a Technofeminist Future for Women and Girls

In her 1986 commencement speech to the women graduates of Bryn Mawr University, Ursula Le Guin gave her audience a call to action:

> Now this is what I want: I want to hear your judgments. I am sick of the silence of women. I want to hear you speaking all the languages, offering your experience as your truth, as human truth, talking about working, about making, about unmaking, about eating, about cooking, about feeding, about taking in seed and giving out life, about killing, about feeling, about thinking; about what women do; about what men do; about war, about peace; about who presses the buttons and what buttons get pressed and whether pressing buttons is in the long run a fit occupation for human beings. There's a lot of things I want to hear you talk about. (Le Guin, Commencement Address, n.p.)

Le Guin, who died in January 2018, depicted in her novels, as well as her published nonfiction, sociopolitical futures in which gender was neutral, fluid, and not a predetermination of what either a woman or a man could or must be. I began *Technofeminist Storiographies* with a discussion of dramatizations of the individual genius narrative, notably Alan Turing and Stephen Hawking, who also died in early 2018, his place in the history of scientific discovery as secure as Galileo and Einstein. After numerous dramatizations over the years, Einstein's own life has recently been adapted for the small screen in the form of *Genius*, and as with the 2015 film *Steve Jobs*, is based on a biography (2007) by Walter Issacson, *Einstein: His Life and Universe*. The series is grounded in the German-born physicist's historical and cultural moment, his flight from the rise of the Nazis and emigration to the United States, and his complicated romantic and domestic life.

This book has deployed a technofeminist re-reading of the lives of multiple women innovators, grounding their work in similar historical and cultural moments that both enabled and constrained their individual and their collaborative achievements, and impacted how those achievements have been represented. For Judy Wajcman (2004), "the enormous variability in gendering by place, nationality, class, race, ethnicity, sexuality and generation makes a nuanced exploration of the similarities and differences between and across women's and men's experience of technoscience all the more necessary" (p. 8). Part of that difference involves the historical representation and visibility of that innovation. For that reason, I have contrasted those histories and their representation with those of masculinist archetypes and the resulting mythologies of figures such as Steve Jobs.

Today, our current cultural moment is indeed complicated for women, providing similar possibilities and constraints for their careers. In the spirit of Le Guin's epilogue, women have not been silent. Merriam Webster's (2018) decision to make "feminism" the 2017 word of the year is a fortuitous one, based on a broad range of feminist rhetorical activism in response to the pre- and post-2016 election and its resulting policies that impact not only the diversity of women within and across cultures but other groups whose access to civil liberties is mediated by inequitable systems of difference. This includes, as I have stressed in this project, the perceived role of women in the history of computer culture. In speaking about their rationale, Merriam Webster representatives noted that "No one word can ever encapsulate all the news, events, or stories of a given year—particularly a year with so much news and so many stories. But when a single word is looked up in great volume, and also stands out as one associated with several different important stories, we can learn something about ourselves through the prism of vocabulary" (n.p.).

As an English professor and as a technofeminist, I've been always interested in vocabulary, the language and rhetoric that dominates the master narrative about computers and how that constrains women's technological histories and stories, the ones that haven't been told, resulting in voices and spaces left behind at the margins of contemporary digital culture. My hope is that the women and girls of today, and tomorrow, need not remain silent about their experiences with technology, are guaranteed equal opportunity to pursue technology education and career paths, and that they are as equally aware of women's historical contributions as they are of men's. My purpose in in tracing some, but certainly not all representations of women in the history of information technology, has been to recover these representative stories to show that our understanding of technological innovation needs to be extended beyond the mythological figures of a Steve Jobs. These figures are culturally inscribed as individual heroes, however flawed, at the expense of more collaborative models of technology history. Today's students and future IT workers need to know of the historical contributions of Ada Lovelace, who collaborated on the first computer algorithm with mathematician Charles Babbage, or the actress Hedy Lamarr, who with the composer George Antheil patented the first frequency hopping technologies meant to help the Allied forces in World War II, that are the precursors to the wireless and Bluetooth connectivity we take for granted today. And they need more diverse role models not only in classrooms and workplaces that support the inclusion of gender and racial diversity but also in contemporary new media depictions of that identity. Speaking of the cultural representation of African American women, Safiya Noble (2018) powerfully argues, "just exactly how it can be that Black women and girls continue to have their images and representations assaulted in the new media environments that are not so unfamiliar or dissimilar to old, traditional media depictions" (Kindle edition, n.p.).

NO TIME LIKE THE PRESENT?

In early 2018, we marveled at Tesla's Elon Musk's Car in Space, and recent articles on Musk (Cofield, 2015) often compare Musk simultaneously to comic book billionaire superheroes such as Batman's alter ego Bruce Wayne, or supervillains such as Superman's foil Lex Luthor. Yet it's important to remember, as I chronicled in chapter 2, the "hidden figures" of both book and film fame from the NASA program: Katherine Johnson, Mary Jackson, and Dorothy Vaughan. It is important we know and say their names, just as it is important we know and say the names of those women programmers of the original ENIAC computer, worth repeating again in this project's conclusion: Kathleen McNulty Mauchly Antonelli, Jean Jennings Bartik, Frances Snyder Holberton, Marlyn Wescoff Meltzer, Frances Bilas Spence and Ruth Lichterman Teitelbaum.

Part of the interest in feminism in the early days of the Trump administration involved the increasing exposure of rampant sexual harassment across all forms of business and industry, from the ouster of individuals such as Fox News' host Bill O'Reilly to NBC's Matt Lauer in the news industry, to politicians on both sides of the aisle, Senator Al Franken and the failed U.S. Senate candidacy of Republican Roy Moore, along with the dramatic fall of powerful Hollywood producer Harvey Weinstein. The latter helped give rise to intense social media activism in the form of the #MeToo movement and other forms of hashtag advocacy that I briefly chronicled in chapter 3. Numerous actresses came forward publicly and in some cases anonymously in a startling expose published by Ronan Farrow in 2017, which in April 2018 received the 2018 Pulitzer Prize for service, shared with *New York Times* writers Jodi Kantor and Megan Twohey. In response to the Weinstein scandal, actress Alyssa Milano popularized the use of the #MeToo hashtag on Twitter to express solidarity. As Milano (2017) stated, "Suggested by a friend: 'If all the women who have been sexually harassed or assaulted wrote "Me too." as a status, we might give people a sense of the magnitude of the problem.'" Within 24 hours of Milano's tweet, the hashtag had been posted more than 500,000 times and has fostered conversations across a range of contexts, from the military, to the music industry, to the church, and to sports and sports medicine, notably in this latter context due to the exposure of USA gymnastics doctor and osteopathic physician Larry Nassar at Michigan State University (Hobson & Boren, 2018), for the sexual assault of nearly 300 minors over the course of his career.

Soon after Milano's post, criticisms emerged about the lack of acknowledgement of #MeToo's original use by activist Tanara Burke, who first used the term over a decade ago to make visible sexual abuse of women of color. In reporting on the rise of the term and Burke's ongoing activism, *The Washington Post's* Abby Ohlheiser (2017) documented the recovery of and advocacy for Burke's work for a more diverse group than many of the white women now using the term were aware: "Part of the message here is that a viral hashtag that was largely spread and amplified by white women actually has its origins in a decade of work by a woman of color." Burke has reported to be both gratified and overwhelmed with the support and recognition she has received, including as I noted in chapter 3, part of the 2017 *Time Magazine* Person of Year as one of numerous "Silence Breakers." But in an exclusionary move that warranted the backlash it received, Burke was left off the cover that featured white entertainers Ashley Judd and Taylor Swift, her origin story receiving limited coverage in the cover story. For Sheryl Estrada (2017) the silence about Burke's contribution speaks volumes: "for the Person of the Year issue, the focus of Burke's original endeavor is framed by the one tweet from Milano. A Black woman's quest to change society for the better is now better accepted because a white advocate can also be the face of it." Burke, however, along with figures that included Microsoft CEO Satya Nadella, was named one of *Time*'s 100 Most Influential People of 2018, complete with her own cover.

As I also stressed in chapter 3, #MeToo certainly can apply to Silicon Valley, and in a 2018 presentation with journalist Ronan Farrow at Ohio's Youngstown State University, Burke relayed a conversation she had with Facebook's Sheryl Sandberg, indicating that the female executive expressed dismay with the idea expressed by IT managers that to avoid harassment and sexism that they just won't hire women. With such a sentiment, however based in hearsay, the IT industry inherently reinscribes the exclusion they purport to decry, unwilling to challenge the male mythos that defines computing culture in the both media representation and workforce demographics. In the introduction to his collection *Gender Codes: Why Women are Leaving*

Computing, Thomas J. Misa (2010) contends that both in the academy and the workforce, "all reform efforts need to confront the distinctive culture of computing. If language creates culture, then computing has created its own universe....Popular images in advertisements, movies, computer games, and computer magazines all tend to reinforce the male dominance of the field" (pp. 11-12).

This male dominance has too often enabled attempts to silence and threaten women across computing industries, including game developer Zoe Quinn, as I discuss briefly in chapter 1. Quinn's (2017) memoir, *Crash Override: How Gamergate (Nearly) Destroyed My Life, and How We Can Win the Fight Against Online Hate*, details her individual harassment initially by an ex-boyfriend and ultimately a global network of male perpetrators of physical and emotional violence. Quinn's story and her willingness to talk and fight back has led to the Crash Override Network, a resource site for anyone experiencing similar online abuse, something women experience in many contexts, from revenge porn sites to stalking and surveillance, instances where the very technologies women had a historical and contemporary role in developing are used to oppress and silence them. Here, I would be remiss if I didn't mention Bailey Poland's (2016) useful overview of the online culture of cybersexism in her book *Haters: Harassment, Abuse, and Violence Online*, and her discussion of the multiple ways individual women and various activist groups are fighting back.

One such activist example that bridges the academic and the popular is Sam Blackmon and Alex Layne's development of the website Not Your Mama's Gamer (n.d.), "designed as a space that would bring scholarly endeavors in line with personal passion, a space that would combine feminist interrogation of games with the games community." The site features podcasts with more diverse gamer designers and equally diverse games. And although it has recently transitioned into a more formalized online academic journal, *Not Your Mama's Gamer* (NYMG) has nonetheless represented an activism that involved graduate students to foster both feminist critique and active visible participation in gaming culture. Given #gamergate, the Quinn and Sarkeesian cases, and the "epidemic" proportions of online violence that Poland exposes, these activist responses are clearly vital. Nevertheless, for Poland, "Solving online abuse will require giving voice to the women who are often ignored by those who have the opportunity to report on Internet harassment, as well as relying on their experiences and expertise as guideposts to a better and safer online environment" (p. 252).

Quinn's experience aligns with Misa's point that computing culture is not just restricted to one arena or one type of job role but encompasses multiple industries, supporting my focus in this project on discussions of film, comics, games, and women's rhetorical representation, or lack thereof, in these media artifacts. Misa also asserts that "effective interventions to improve professional practices in computing (and other technical fields) require greater historical awareness" (p. 13). He inevitably calls for us to go back decades to better understand the evolution of computer culture and computer science as a discipline; however, even he stops short of acknowledging the legacy of women whose innovations shaped the larger social network as we experience it today. That such recognition is needed is evident any time we read any origin story of networked computing that doesn't include, for example, the collaborative innovations of Hedy Lamarr and George Antheil, as we see in the website for the Wifi analytics firm Purple. In 2014, Purple Marketing Manager Becki Wood published a blog post about the history of Wifi that featured a graphic, appropriately purple-hued, depicting the evolution of a man from early primates to homo sapiens. Wood writes "But how many of us know the full history behind WiFi technology? When was WiFi invented? How exactly does it work? And just how far it has come in 20 years? Here we've explored the history of WiFi, from where it began, what it has helped us achieve, and what future it promises us as we become increasingly interconnected." Wood's narrow window for the "history" of Wifi technology focuses on the potential of Wifi to help customers "gain thorough amounts of user data," concluding that Wifi "has become an essential part of our personal and professional day-to-day, and is constantly improving our efficiency, our communication, and is persistently encouraging the technology industry to push the boundaries of what's possible."

Yet numerous tech-based companies and their titanic leaders have come under fire for extending "what's possible" to include the use of analytics and resulting algorithmic marketing of products and, in the most recent presidential campaign ideologies, the marketing of political ideologies. A notable example is Cambridge Analytica's misuse of data from approximately 50 million Facebook users (Rosenberg, et al., 2018) to manipulate content to influence the 2016 US presidential election in favor of then candidate Donald Trump. The scandal has led to requests by the US Congress and the British Parliament for CEO Mark Zuckerberg to appear before them and provide evidence. Throughout this project, I have chronicled the cultural and historical moments that enabled Hedy Lamarr, Grace Hopper and the women of ENIAC to contribute positively to the technologies that govern our professional and social lives today, including the Second World War. Undoubtedly, such exigencies are in stark contrast to the entrepreneurial exigencies for a Steve Jobs, Mark Zuckerberg, and the many individuals working in today's IT industry, and admittedly the scandal surrounding Facebook privacy breaches is certain to disrupt Zuckerberg's attempt, as I discussed in chapter 3, at a more philanthropic legacy compared to his longstanding role model Bill Gates. Although I am not attempting to reinscribe essentialist views of gender in noting Zuckerberg's ethical challenges, a portion of Le Guin's commencement address that begins this chapter is worth repeating about speaking out "*about what women do; about what men do; about war, about peace; about who presses the buttons and what buttons get pressed and whether pressing buttons is in the long run a fit occupation for human beings.*" What may get lost in the desire for profit within and outside Silicon Valley culture is the connection between functional, critical, rhetorical, and ethical literacies involving technology and its representation in the larger culture.

TECHNOFEMINIST RE-CODES

As I argued in chapter 4, while there are notable examples to foster more gender-fair depictions of women's and girls' relationship to technology in a counter to such artifacts as *Barbie: I can be a Computer Engineer*, there is more work to do to counter the focus on what, how, and why girls do versus what boys do in our culture in general and within tech culture as well. How might the development of technological skills enable women and girls to speak back to cultural assumptions about gender, including Barbie. A notable example of such "back-talk" occurs in the work of Juliet Davis, an artist at the University of Tampa and producer of the online game Polystyrene Dream. As a form of social satire, the game features the ageless Barbie in a series of cultural and social contexts, from the military, the church, to even the presidency, clearly this latter scenario not yet within reach. One specific tab for "Barbie's Fertility Clinic" provides the user options to create a different version of Barbie and Ken, from Handicapped Barbie, Afro Barbie, and Fat Barbie, to Trans Ken or gay Ken and Ben. All of these models are "discontinued" in favor of the one right answer: Miss America Barbie, complete with pink cape and crown. Davis' goals are notably activist: "Our four-decade fascination with the Barbie doll holds unsettling implications about how we view ourselves and our world," and our emphasis on "physical perfection."

Models like Gene Luen Yang's *Secret Coder* series that I feature in chapter 4 represent an attempt to disrupt such narratives that limit women and girls. Despite the important work to disrupt the pink aisle across media genres there is still a long way to go, as several VTech toys aimed at girls, including the digital camera Kidizoom, and a smart pink-encased faux smart phone, Kidibuzz, represent. Similar to the bookstore display I shared in chapter 1, the Kidizoom packaging encourages girls between ages three to eight to "explore your creativity," even as it promotes the "selfie mode" above all the other camera features. Although some research has resulted in calls to make the study of computer science for girls less stereotypically "geeky," reinforcing sexist assumptions about who uses technology and how is not the way to improve attitudes or aptitude toward technology use.

As the site for the advocacy group Let Toys Be Toys posted in late 2017, "gender bias means STEM toys are often targeted squarely at boys, (or else given a dusting of pink glitter and lipstick as if that's the only way to get girls interested)." These concerns extend beyond toys to the other media I have featured throughout this book. In focusing on bias in children's books, for instance, Let Toys Be Toys questions how even the lack of female villains in books and other children's media represents the cultural assumption about the limits on girls' ambitions and intelligence. When such villains do exist, their role is typically limited by a desire to vanquish a rival, usually a princess or a queen. Even the *Minions* movie infamous female supervillain Scarlett OverKill (voiced by actor Sandra Bullock) is ultimately undone by her unquenchable desire for the jeweled crown of the Queen of England so that she too can achieve her lifelong dream of being a princess. Ironically, Bullock advocates the need for female villain role models in the 2018 *Oceans 8*. As the indomitable Debbie Ocean, she motivates her team by stressing their status as role models: "Somewhere out there is an eight-year-old girl lying in bed, dreaming of being a criminal. Let's do this for her." Although this statement applies to the plot's criminal theme, it can apply to the diversity of roles for women in film and the ability to remix the bromance of the Ocean's franchise in general, a feminist statement that seems better received than earlier attempts, such as the all-women Ghostbusters remake of 2016 (Sims, 2016) that I reference in chapter 1.

One response to the need to transform cultural assumptions is through girls' technological literacy development, developed in the spirit of Grace Hopper's original call for educational initiatives and from communities such as "About Feminism." As an example of such technology education, for five years, I co-developed and directed The Digital Mirror Computer Camp for Girls, a four-day residential computer camp for middle-school adolescent girls, the very population the American Association of University Women identified, as I noted in chapter 1, as being vulnerable to cultural assumptions about who uses technology and how, including within classroom contexts. Based on that connection, the camp was partially funded through a national AAUW Community Action Grant. The camp responded to these concerns by relying on only women facilitators, including graduate students in my university's doctoral program in Rhetoric and Writing, to help deliver curriculum in web-based and social-media authoring, along with digital image, video, and audio editing. These girls had no idea who Ada Lovelace, Grace Hopper, Katherine Johnson, Mary Jackson, and other women innovators even were. To better represent the role of women working in technology, we hosted female speakers from Digital Art, Computer Science, and Educational Technology, talking about their own education and career trajectories. Our shared curricular goal was to move beyond the mere emphasis on functional literacy to instead foreground the critical and rhetorical aspect of digital literacy in providing opportunities for the twenty to twenty-five adolescents annually enrolled in the camp. This included time, through blogging and video production, to reflect on their reasons for using or not various digital media tools, and then to develop a portfolio of artifacts that allowed them to share these reflections (hence the mirror theme of the camp) in a showcase section with parents and family members on the camp's final day. These forums were more than celebratory; they also helped to show parents, often reluctant to have their daughters work in online spaces based on larger narratives about safety and privacy, that it was important to provide time and space for experimentation and play with digital identity formation and the important role of technology in girls' literate lives.

Although the interest among parents was strong as I began to promote the camp in local schools and among faculty colleagues, I frequently received questions about why an English department faculty member was coordinating such an initiative; why not computer science faculty, for instance? The response was and is simple: Being literate in the twenty-first century requires an emphasis on a range of multimodal tools that represent the way students, workers, and citizens will and must communicate in not only functional but also critical, rhetorical, and ethical ways. This makes the role of rhetoric specialists as vital to this activist, educational goal as computer scientist and mathematicians. While *Newsweek's* 1975 "Why Johnny Can't Write" articulated a literacy crisis in which the lack of writing skills on the part of college students before they arrive on campus, while they are there, and after they graduate "may not even qualify them for secretarial or clerical work" (p. 58), it was clear that this crisis, real or imagined, should be our collective concern as a society. Writing and other communicative acts have always been technological, and thus from a technofeminist standpoint, it is now imperative to ask which groups have access to those technological literacies and why that should also be our collective concern as a society, and across academic disciplines, as the Digital Mirror camp represents.

There have been numerous articles and blogs devoted to "Why Johnny Can't Code" or "Why Johnny Can't Code Good," calling for students to move beyond functional literacies of reading and writing to rhetorical literacies of technology creation, something tech industry journalists have consistently recommended. Indeed, Paul Roberts's 2013 incarnation of "Why Johnny Can't Code" advocates moving toward producing "students who aren't merely technology consumers ('I buy Angry Birds and play it on my iPad') but technology creators ('I wrote a game that's better than Angry Birds, and sold it on the AppStore')." Writing studies scholar Annette Vee (2017) similarly argues for an understanding of code as a literacy with both material, rhetorical, and ultimately intersectional factors and must be accessible in a broader range of school-based learning contexts, given the inequities Vee notes in extracurricular programming. For Vee, given the undeniable role of code in our access to information and communication venues, code is in fact a "platform literacy," no longer a specialized skill but rather a basic one that can be deployed in public activist contexts in which we participate as citizens in the twenty-first century (p. 209). But instead of asking "Why Johnny Can't Code," we should ask "Why Sally, or Maria, or Toneisha, or Yasmin Can't Code and Why?"

TECHNOFEMINIST BACK-TALK

Regardless of who's doing the coding and why, girls need the chance to develop, as Stuart Selber (2004) has outlined, functional, critical, and rhetorical literacies, moving beyond the ability to use and consume technology but produce persuasive responses to their culture as engaged literate citizens. In light of the abuses of technology to harass others and spread rhetorics of hate, there is a need for ethical literacies as well. They also need to use these and other tools to reflect on their sense of self and society, and to inevitably gain the power to speak back from within and outside that culture. In this way, technological literacy becomes activist, something documented in the "About Feminism" and "Table Flip" movements I highlighted in chapter 1, but also something beyond the IT industry, as we see with the #MeToo movements and the other forms of hashtag advocacy.

My rationale for discussing the Digital Mirror Camp is also about what the participation of graduate students in such an initiative enabled them to do on their campuses as future faculty. I would argue that this type of sustainability is something that sustains not just a more qualitative research emphasis but also technofeminist coalition building for women participating in virtual spaces, both inside and outside the academy. Ultimately, the collaborative efforts with the graduate student women with whom I was honored to work on the Digital Mirror Computer Camp, who served as mentors to the girls themselves, helped to model that technology wasn't just a "guy thing." The importance of mentoring also became evident over time, for although the Digital Mirror Camp ended after five years, a number of the former graduate students went on to develop similar camps at their future institutions, including Dr. Jen Almjeld's Girlhood Remixed at New Mexico State University. This camp placed "more emphasis on identity construction but retained the focus on gender constraints in online spaces. The camp was built on the best practices of the Digital Mirror while reflecting Almjeld's scholarly interests in girlhood as a construct and identity with and through technologies" (Almjeld & England, 2016). Despite Almjeld's departure to a new academic institution, the Girlhood Remixed camp continued, similarly through the same form of graduate student mentoring when doctoral candidate Jen England assumed leadership as camp director, maintaining the camp's outcomes: "the development of technological, critical, and rhetorical literacies as the girls pursued their own technology-related goals; and the crafting of powerful, positive articulations of girlhood through the girls' production of new media and technologies" (England & Cannella, 2018, p. 75).

Based on the longstanding emphasis on community literacy and service learning within the disciplines of women's studies and writing studies, such projects represent a form of technofeminist activism that is an emerging subgenre of civic engagement and political activism. This results in research that broadens our understanding of the material and cultural conditions that impact digital literacy acquisition both inside and outside the academy, including for girls in STEM. The AAUW has evolved its advocacy for gender equity to include activist, educational training for girls that include camps such as Tech Trek, a one-week camp that like Digital Mirror, had its origins in a community action grant that has now grown to over 20 camps across the country.

In addition to my own technology camp and those of my former students now making an impact on their own campuses and in the larger profession, digital rhetoric scholars such as Mary Sheridan (2018) and her colleagues and graduate students at the University of Louisville have engaged in similar technofeminist practices through their work on the Digital Media Academy. The camp, a two-week summer day camp for middle school girls, is designed to help girls to develop the "technological, critical, and design literacies they needed to create digital messages of their own choosing, thus encouraging girls to be critical producers, not just consumers, of digital media" (p. 222). But as with the Digital Mirror Camp, Sheridan's Digital Media Academy was also designed both to help the camp participants "recognize and redesign the pervasive sexualized and commercialized images of what it means to be a girl today"(p. 222) and to help the graduate students again develop the civic, pedagogical, and technofeminist dispositions that inform activist initiatives and speak back to the gendered rhetorics that constrain women's and girls' technological aptitudes and attitudes. These local initiatives align with larger national ones such as ChickTech, which works with companies to "create a more empowered and inclusive workspace," and stresses that "a crucial aspect of all our programs is finding women who would not otherwise have thought of themselves as capable of succeeding in a high tech career." ChickTech also confronts what I refer to as the Princess Problem in chapter 4 on its "About" page:

> Like many in this generation, we are reclaiming the word "chick." As women, we have the right to call ourselves by whatever label we want–whether that's "girls," "women," "female," . . . or "chick." We've seen many instances of people taking back words that are used when women are demonstrating more traditionally masculine words, which we love and support. However, we've

found that when women choose to reclaim and use a word (or color) that is more feminine, this continues to be viewed negatively by our society. At ChickTech, we believe that as a woman, you can be whoever you want to be . . . including a feminine woman . . . and still be technical. We feel that when women and girls do not have the power to call themselves what they want, they are being disempowered and forced to fit stereotypes of who can be taken seriously and who can't. We will continue to take back the word "chick" by empowering and supporting powerful, confident, and creative women and girls! (ChickTech, n.p.)

Although the site sports at least one pink heart on its donate icon and a pink flower as its URL icon, an equally dominant graphic on the ChickTech site features a woman of color in comic form, riding a motorcycle in a futuristic, superhero motif, with the accompanying caption "Accelerating Tech with Diversity," a stark contrast to the CodeBabes website I discuss in chapter 1. Thus for ChickTech, technology education has as much to do with attitude as aptitude and has to include employers. Along with Girls Who Code and other more diverse initiatives that include Black Girls Code, these initiatives represent for Vee an opportunity through programming to "seize power that has historically been wielded against them" by "flipping the dominant script about coding" (p. 79). These support networks extend to LGBT tech groups such as OutinTech, which provides training and internship opportunities for LGBT youth to foster further interest in pursuing tech careers; and Maven Youth, which includes among its programs a "Tech Give Back" initative where tech companies, as their website indictes, "donate used employee laptops to place into the hands of the next generation of queer tech innovators!" (mavenyouth.org). All of these initiatives help to tell a different story of innovation, that computer culture need not remain predominantly white, male, and straight.

Given the Facebook-Cambridge Analytica controversy, we may be questioning our own privacy and security on Zuckerberg's platform, yet it is nevertheless important to concede such tools can help spread the word about historical and contemporary contributions to STEM education. For instance, The National Women's History Museum's (NWHM) Facebook Page published the following as part of its #OnThisDay hashtag:

#OnThisDay in 1916, a newly published book encouraged girls to build electromagnets, study the aerodynamics of flight, and send messages using Morse code. It instructed girls in the mechanics of pitching a tent, building a campfire, and using a compass. In a society that adhered to Victorian beliefs that there were "boy" activities and "girl" activities, emboldening girls to become knowledgeable and proficient in non-traditionally feminine skills was somewhat radical. What was the title of this revolutionary publication? *How Girls Can Help Their Country: A Handbook for Girl Scouts.* The author was

Juliette Gordon Low, the founder of the Girl Scouts of the USA. (NWHM, n.p.)

With this point in mind, it is equally fitting that I conclude this project during the last days of Women's History Month in March 2018. The 2017 Women's March on Washington evolved from a Facebook page created by Hawaii grandmother Teresa Shook and through other consolidated grassroots social media campaigns to over six hundred marches on seven continents. The event was a powerful assertion of, à la Hillary Clinton's 1995 speech to the United Nations Fourth World Conference on Women, "Women's rights are human rights and human rights are women's rights." Despite the power that the twenty-first-century Women's March represents in the post-inaugural response to the election of Donald Trump, we are not so far removed from the struggles of our early suffragist sisters. It is important to recall the efforts of suffragists Alice Paul (1885–1977), Lucy Burns (1879–1966), and the National Women's Party, who organized the group of Silent Sentinels, women who picketed the White House in 1917, the first activist group to do so, just over 100 years ago. The full circle of a century represents an important opportunity to reflect on the power of feminist—and I would argue technofeminist—rhetorics past and present to make a difference. In the context of the history of information technology, both women and men must have similar opportunities to learn and honor the role of women in that history in order to reshape the pervasive rhetoric that computing culture's past, present, and future will continue to be male.

CODA

As Le Guin reminds us, gender neutrality must not be the stuff of science-fiction and fantasy, and the celebration of women's contributions in all aspects of scientific and technological innovation must be the norm, not the exception we label as long overdue. Even as I make such a call, it is equally important to acknowledge the limits of the historical moments I've included in this project; they undoubtedly represent a western European history of innovation given the origin stories of computing culture in Great Britain and the United States, not to mention our continued demographics that position both women and people of color as a minority workforce in the American IT industry. In late 2017, the *Los Angeles Times* received almost as vocal an amount of criticism as the *Time Magazine* Silence Breakers Person of the Year cover, in this case for its "Shift in Focus" cover story featuring six white actresses, diverse in generation but not in skin or hair color (all, except for actress Jessica Chastain, decidedly blond), calling for "a change in the way many stories are told." The social media backlash through Twitter was intense (Weingarten, 2017), with numerous tweets connecting the privileging of white women's stories with the themes of Jordan Peele's 2017 film *Get Out* and its extreme horrific depiction of blasé white privilege among upper-middle class suburbanites. Even as we celebrate more gender balance in both biographic and comic book dramatizations such as *Wonder Woman* and more recently *Black Panther*, we must foster stories that promote a diversity of experience and representation.

A proactive response to avoiding gender stereotypes with a complicated character like Wonder Woman, as discussed in chapter 4, includes Chicago-based artists and twin sisters Catherine and Sarah Satrun's (Satrun, 2013) "WE Are All Wonderwomen" poster, which features women of different ethnicity and body types, along with differences in ability via the inclusion of a woman in a wheelchair. And despite these economic realities facing black women who tech, as I've highlighted throughout this book, several contemporary media depictions of black women and technology are more positive. These include the film and comic representations of Shuri, the princess in the fictional kingdom of Wankanda in Marvel's *Black Panther* series. Shuri is on equally powerful footing to her brother, King T'Challa, whom she is able to help through her advanced technological prowess. Princess Shuri is a clear alternative to the princess images presented to adolescent girls, and her role in the film *Black Panther* is that of healer of men. In the larger *Black Panther* series, Shuri will become the ruler herself after T'Challa's battle injuries and overall represents a model of mind-body superpower that disrupts the male "titans" of both comic and Silicon Valley culture. Also notable in the film version of Black Panther is Shuri's collaboration with Nakia, another powerful female character often depicted as a villain but an important part of a triumvirate that includes both Shuri and the female general Okoye. In her *bitchmedia* article "The Women of Wakanda," Evette Dionne (2018) asserts that

> Okoye, Shuri, and Nakia inherently understand that Wakanda is bigger than their individual contributions, and they can only retain their sovereignty if they work together. These women are *Black Panther*'s true revolutionaries. For all of the greatness that Marvel has produced, it has never figured out how to present well-developed, multidimensional, and compelling female characters who don't exist to serve the arc of male characters. (Dionne, n.p.)

In this way, the women of *Black Panther* combat the forces of evil together, representing a model that mirrors women's historically collaborative contributions to information technology, including Ada Lovelace and Charles Babbage, the six ENIAC programmers, Hedy Lamarr and George Antheil, the African American women computers of NACA/NASA, Mary Jackson and Kazimierz Czarnecki, and Grace Hopper and her various teams. Just as this project and others recover and make visible these women-centered or gender-fair collaborations throughout information technology history, it is just as important to acknowledge the need for more diverse representations as well, to not presume that simply to focus on gender is enough to create a historiography that accounts for race, class, sexuality, and other intersectional identity formations.

Nevertheless, there are a range of media texts and contexts that highlight the place of women in the history of information technology. In her book *Broad Band: The Untold Story of Women Who Made the Internet*, Claire Evans (2018) details the long list of women who are as responsible for the digital age as Bill Gates, Steve Jobs, and Mark Zuckerberg and who are largely unknown or unacknowledged. Moreover, Evans distinguishes between the types of goals historical and contemporary figures have in their roles as innovators:

> There are technical women in these pages, some of the brightest programmers and engineers in the history of the medium. There are academics and hackers. And there are culture workers, too, pixel pushers and game designers. . . . Wide as their experiences are, they've all got one thing in common. They all care deeply about the user. The are never so seduced by the box that they forget why it's there: to enrich human life. If you're looking for women in the history of technology, look first where it makes life better, easier, and more connected. (Evans, p. 111)

Evans includes both Ada Lovelace and Grace Hopper in her engaging yet cohesive narrative, grounding each history in their kairotic moments in time. Speaking of Lovelace, Evans concludes "she could have done so much more, and it's evident that she wanted to. Many brilliant women—born in the wrong centuries, the wrong places, or hoping to make an impact on the wrong field—have suffered similar fates, and far worse" (p. 22). If Vee's contention is that we should all regard code as a platform literacy as basic as alphabetic reading and writing, then we must ensure the women coders, including Lovelace, are part of a "platform history," her name and others as known to today's software engineers as the male visionaries that they have been enculturated to emulate, for better or worse. This visibility must span both academic and professional contexts, something I have attempted to do in this project by drawing upon a diverse array of scholarly, popular, and industry sources.

In 2017, the *New York Times* appointed Jessica Bennett as Gender Editor, designed in part to grow women's readership of the 167-year-old periodical but also to rectify a present absence of women's stories. As Bennett writes,

> I see gender as a lens through which we view global storytelling. So that certainly means writing about feminism and women's roles in politics and culture and economics, but it also means covering masculinity and sexuality

and gender fluidity and race and class and looking at science and health and parenting and sport all through this lens. (*New York Times*, n.p.)

Soon after Bennett's appointment, Amisha Padnani was named the *Times* Obituary Section Digital Editor. Padmani soon partnered with Bennet to produce "OverLooked," a digital series honoring the women who had not received a *Times* obituary. For Padnani (2018), "as a woman of color, I am pained when the powerful stories of incredible women and minorities are not brought to light." "Overlooked" (2018) begins with the simple statement, "Since 1851, obituaries in the *New York Times* have been dominated by white men. Now we're adding the stories of remarkable women." Included among that list is Ada Lovelace. And in addition to the original obituaries of ENIAC programmers Jean Jennings Bartik in 2011 that I discussed in chapter 2 and Frances Holberton in 2001 (Lohr, 2001), Hedy Lamarr's status as an entertainer warranted her 2000 *Times* obituary. Yet Richard Sevros's *Times* memorial, "Lamarr, Sultry Star Who Reigned in Hollywood of 30s and 40s, Dies at 86," mostly reinforces the larger pop culture assessment of Lamarr as a delusional divorcee multiple times over, and in her later years, as a faded and litigious star of the silver screen. It also casts doubt on Lamarr's contributions to the patented technology on which she and Antheil collaborated.

And while Hedy Lamarr's documentary *Bombshell* has been screened at countless arts centers, university theatres, and museums, the general public's introduction to this resurgence of Lamarr's story includes a March 28, *People Magazine* Story "Who Was Hedy Lamarr? All About the Tragic Film Star and Secret Inventor in Tonight's *Timeless*. The film's producer Alexandra Dean is quoted, noting the obscurity of Hedy's story: "It's such a current story in a time when only about 24 percent of the workforce in STEM fields are women. We hear again and again that women don't have any role models and there's nobody who came before us, but what if people came before us but their stories weren't known, like Hedy?" Nevertheless, *People*'s story foregrounds the tragedy of Lamarr's reclusive final years in which she mostly hid to avoid the public loss of her famed beauty, despite the fact that, as Dean notes in the article "Almost everybody in the developed world uses Hedy's technology today." Meanwhile, in the quarter-century since her death in 1992, Grace Hopper's legacy has fared much better in terms of media representation. In addition to her posthumous Presidential Medal of Freedom Award in 2015, Google announced in February 2018 that it will back a big-screen biopic based on Kurt Beyer's book that proved so helpful to me in chapter 2; Beyer will serve as the film's consultant. As background, several online technews and film sites (N'Duka, 2018) quote Google's head of Women in Media Strategy, Courtney McCarthy: "Google's research shows that perceptions matter when motivating women to pursue Computer Science, and we're excited to have Grace's story finally told to inspire a new wave of technologists."

Whether that inspiration takes hold, and the film is actually made, remains to be seen, given the material and cultural conditions that mediate women's current and future educational and career path. That we must remain vigilant is evident on a daily basis, as news and social media report on both the educational the tech industry trends that in some cases empower and in others continue to disenfranchise women and girls. As recently as April 2018, the University of Maryland's Computer Science Program faced backlash over its disparate messaging to female vs. male graduate students on its website, warning women that "your students may have difficulty accepting you fully in a scientific field which they may, for whatever reasons, associate with male activity. . . . It's unfortunately the kind of practice you're going to need in the future; students may not be the only ones who have difficulty accepting you as a professional." Before the university could remove the material, which also included directives to male students to beware those female students who "may attempt to capitalize on the male-female dynamic to their own advantage," computer science student Annie Bao screen captured the information and posted it to Twitter (CBS Baltimore, 2018), where it has been retweeted and also shared widely on Facebook. Although material from the handbook was dated, Bao and others were outraged that in a culture of #MeToo and #TimesUp that the department was not more aware of the sexist values it represents, values they claimed to refute in its apology.

One positive outcome of the Maryland case is the way future generations of women computer scientists publicly spoke back to these gendered rhetorics of technology; as Bao explained in the news story, "We need to work toward a change and a better future." Similar goals are at play in the 2018 book *Women and Ideas in Engineering: Twelve Stories from Illinois*, which as the title suggests features the role of women in the engineering program at the University of Illinois and also directly talks back to male-dominated histories, in this case the 1967 book *Men and Ideas in Engineering: Twelve Stories from Illinois*. As I have stressed in this project, the efforts to document the role of women in the history of information technology is vital to both STEM education initiatives that positively impact the future of the IT industry. To borrow from another group of celebrity women organizing online and off against sexual harassment and assault (not to mention the earlier assessment by Hedy Lamarr to her belated accolades with which I concluded chapter 3), "Time's Up," something that the students at the University of Maryland deployed in response to the Computer Science Department website. It's time to speak up and out about women's individual and collective stories of technological innovation and for a larger group of individuals to listen, learn, and share. As Le Guin's epigraph concludes, "There's a lot I want to hear you talk about." I sincerely hope *Technofeminist Storiographies* helps to continue that conversation.

Bibliography

Abbate, J. (2012). *Recoding gender: Women's changing participation in computing.* Cambridge, Mass: MIT Press.

Abramson, A. (28 Feb. 2017). "Hidden Figures" inspiration Katherine Johnson is now part of a Lego set. *Fortune.* Retrieved from http://fortune.com/2017/02/28/hidden-figures-lego-katherine-johnson/

Almjeld, J., & England, J. (2016). Training technofeminists: A field guide to the art of girls' tech camps. *Computers and Composition Online.* Retrieved from http://cconlinejournal.org/fall15/almjeld_england/index.html

Almond, S. (2018). The disturbing evolution of "lock her up." WBUR. Retrieved from http://www.wbur.org/cognoscenti/2018/07/25/lock-her-up-jeff-sessions-steve-almond

Antheil, G. (1940). *The shape of war to come.* New York: Longmans, Green, and Co.

Amanpour, C. (2011). Interview with Bill Gates. "ABC News This Week." Retrieved from http://abcnews.go.com/Politics/bill-gates-dismisses-criticism-steve-jobs-biography/story?id=14845131

American Association of University Women. (2000). *Tech-savvy: Educating girls in the new computer age.* Washington, DC: American Association of University Women Educational Foundation.

AnitaB.org. (2017). Profiles: Meet Brenda Darden Wilkerson, AnitaB.org president and CEO. Retrieved from https://anitab.org/profiles/meet-new-ceo-brenda-darden-wilkerson/

Anthiel, G. (1945). *Bad boy of music.* New York: Doubleday.

Associated Press. (14 February 2017). Admiral's name replaces Calhoun's at Yale. *Diverse Military.* Retrieved from http://diversemilitary.net/2017/02/14/admirals-name-replaces-calhouns-at -yale/?utm_campaign=MILI1702%20DIVERSE%20MILITARY%20FEB15-FINAL&utm_medium=email&utm_source=Eloqua

Asseyev, T., et al. (Producer), & Ritt, M. (Director). (1979). *Norma Rae* (Motion Picture). United States: 20th-Century Fox.

Bahadur, N. (2014). "Dear Kate" ad features woman tech execs in their underwear. *Huffington Post.* Retrieved from https://www.huffingtonpost.com/2014/08/26/dear-kate-ad-tech-execs-women_n_5710999.html

Barthes, R., & Lavers, A. (1972). *Mythologies.* New York: Hill and Wang.

BBC Horizon. (1964). Arthur C Clarke imagines life in the year 2000. YouTube. Retrieved from https://www.youtube.com/watch?v=q1teW1bmQUs.

Beck, E., Blair, K., & Grohowski, M. (2015). Gendered labor: The work of feminist digital practice. *Kairos: A Journal of Rhetoric, Technology, and Pedagogy*, 20(1). Retrieved from http://kairos.technorhetoric.net/20.1/reviews/blair-et-al/index.html.

Berg, R.A., Schillinger, R.D., & Kingery, E.H. (1967). *Men and ideas in engineering: Twelve stories from Illinois*. Urbana, IL: University of Illinois Press.
Besson-Silla, V. (Producer), & Bresson, L. (Director). (2014). *Lucy* (Motion Picture). United States: Universal.
Berman, P. (Producer), & Leonard, R. (Director). (1942). *Ziegfeld girl* (Motion Picture). United States: Metro-Goldwyn-Mayer.
Bevan, T., et al. (Producer), & Marsh, J. (Director). (2014). *The theory of everything* (Motion Picture). United Kingdom: Universal Pictures.
Beyer, K. (2009). *Grace Hopper and the invention of the information age*. Cambridge, MA: MIT Press.
blackcomputeHer.org. (n.d.) Home: Who we are. Retrieved from https://blackcomputeher.org
Blackmon, Samantha. (n.d.). Not your mama's gamer. Retrieved from http://www.nymgamer.com
Blair, K., & Tulley, C. (2007). "Whose research is it, anyway?" The challenge of deploying feminist methodologies in technological spaces. In D. DeVoss & H. McKee (Eds.), *Digital writing research: Technologies, methodologies, and ethical issues* (pp. 303-17). Cresskill, NJ: Hampton Press.
Blick, H. (Writer & Director). *The Honourable Woman* (Television MiniSeries). United Kingdom/United States: BBC Worldwide. https://www.youtube.com/watch?v=q1teW1bmQUs.
Boyle, D., et al. (Producer), & Boyle, D. (Director). (2015). *Steve Jobs* (Motion Picture). United States: Universal Pictures.
Brockwell, H. (2017). Sorry, Google memo man: Women were in tech long before you. *The Guardian*. Retrieved from https://www.theguardian.com/commentisfree/2017/aug/09/google-memo-man-women-tech-original-computer-programmers
Brown, J., et al. (Producer), Glatzer, R., & Westmoreland, W. (Director). (2014). *Still Alice* (Motion Picture). United States: Sony Pictures.
Bump, P. (2018). Trump's rationalization for calling women "dogs" helped define his campaign. *The Washington Post*. Retrieved from https://www.washingtonpost.com/news/politics/wp/2018/08/14/trumps-rationalization-for-calling-women-dogs-helped-define-his-campaign/?utm_term=.2af65847fba5
Burke, T., & Farrow, R. (28 Mar. 2018). Youngstown State University Centofanti Symposium.
Burleigh, N. (2015). What Silicon Valley thinks of women. *Newsweek*. Retrieved from http://www.newsweek.com/2015/02/06/what-silicon-valley-thinks-women-302821.html
Burrell, I. (2014). Stephen Hawking's wife Jane Wilde on their marriage breakdown: "The family were left behind." *The Independent*. Retrieved from http://www.independent.co.uk/news/people/stephen-hawkings-wife-on-their-marriage-breakdown-the-family-were-left-behind-9949588.html
Butler, J. (1990). *Gender trouble: Feminism and the subversion of identity*. New York: Routledge.
Butler, W. (2014). "Silicon Valley" and responsible satire. *Los Angeles Review of Books*. Retrieved from https://lareviewofbooks.org/article/silicon-valley-responsible-satire
Buzzfeed. (2016). Twitter permanently suspends conservative writer Milo Yiannopoulos. Buzzfeed News. Retrieved from https://www.buzzfeednews.com/article/charliewarzel/twitter-just-permanently-suspended-conservative-writer-milo
Calendrelli, E., & Kurilla, R. (2017). *Ada Lace sees Red*. New York: Simon & Schuster.
Calkhoven, L., & Petersen, A. (2016). *You should meet: Women who launched the computer age*. New York: Simon & Schuster.
Cao, S. (2018). New York is better than Silicon Valley for women entrepreneurs. *Observer*. Retrieved from http://observer.com/2018/01/new-york-silicon-valley-women-entrepreneurs/
Care2 Petitions. (2016). Reconsider the choice of Wonder Woman as the UN's Honorary Ambassador for the Empowerment of Women and Girls. Retrieved from https://www.thepetitionsite.com/741/288/432/reconsider-the-choice-of-honorary-ambassador-for-the-empowerment-of-women-and-girls/
Carly for America. (14 Sept. 2015). Faces. Retrieved from https://www.youtube.com/watch?v=ODfUOnw2x0g

Bibliography

CBS Baltimore. (2018). UMD handbook tells female TAs to expect "challenging behavior," "be patient." Retrieved from http://baltimore.cbslocal.com/2018/04/18/umd-handbook-tells-female-tas-to-expect-challenging-behavior-be-patient/

Chang, E. (2018). *Brotopia: Breaking up the boys' club of Silicon Valley*. New York, NY: Portfolio/Penguin.

Chapman, C. (2018). When it comes to tech, lesbians want to "reclaim the space." NBC News. Retrieved from https://www.nbcnews.com/feature/nbc-out/when-it-comes-tech-lesbians-are-trying-reclaim-space-n853851

Cheddar. (2018). Former Snap employee questions company's culture and diversity. Retrieved from https://cheddar.com/videos/inside-snap-employee-concerns-raise-questions-about-culture-and-diversity

ChickTech (n.d.). Retrieved from https://chicktech.org

Chozik, A. (2015). Hillary Clinton sketches campaign messages in Silicon Valley. *New York Times*. Retrieved from http://www.nytimes.com/2015/02/25/us/politics/clinton-sketches-campaign-messages-in-silicon-valley.html

Clinton, H.R. (1995). Remarks to the U.N. 4th World Conference on Women Plenary Session. American Rhetoric Top 100 Speeches. Retrieved from http://www.americanrhetoric.com/speeches/hillaryclintonbeijingspeech.htm

CNBC. (2017). Women-only screenings of *Wonder Woman* are causing an uproar. CNBC.com. Retrieved from https://www.cnbc.com/2017/05/26/women-only-showings-of-wonder-woman-at-alamo-drafthouse-cause-uproar.html

Cofield, C. (2015). Stephen Colbert says Elon Musk is either a supervillain or a Superhero. Space.com. Retrieved from https://www.space.com/30505-stephen-colbert-calls-elon-musk-supervillain.html

Colatrella, C. (2001). From *Desk Set* to *The Net*: Women and computing technology in Hollywood films. *Canadian Review of American Studies*, 31(2), 1–14.

Conner, L., et al. (Producers). (2017). *Genius* (Television Series). United States: Imagine/20th Television.

Cook, T. (2011). Apple media advisory. Retrieved from https://www.apple.com/newsroom/2011/10/05Apple-Media-Advisory/

Cook, T. (2014). Tim Cook speaks up. *Bloomberg*. Retrieved from https://www.bloomberg.com/news/articles/2014-10-30/tim-cook-speaks-up

Crenshaw, K. (1989). Demarginalizing the intersection of race and sex: A black feminist critique of antidiscrimination doctrine, feminist theory, and antiracist politics. University of Chicago Legal Forum. University of Chicago Law School. Retrieved from https://philpapers.org/archive/CREDTI.pdf

Crowther, B. (16 May 1957). The screen: *Desk Set*; murder and mayhem in *Garment Jungle*. *The New York Times*. Retrieved from http://www.nytimes.com/movie/review?res=9D03E3D91631E63ABC4E52DFB366838C649EDE

Damore, J. (2017). Google's ideological echo chamber. Retrieved from https://assets.documentcloud.org/documents/3914586/Googles-Ideological-Echo-Chamber.pdf

Davis, J. (n.d.) Polystyrene dream. Retrieved from http://www.julietdavis.com/studio/barbie.html

Day, M. (2018). "I felt so alone": What women at Microsoft face, and why many leave. *Seattle Times*. Retrieved from https://www.seattletimes.com/business/microsoft/i-felt-so-alone-what-women-at-microsoft-face-and-why-many-leave/

Deeley, M. (Producer), & Scott, R. (Director). (1982). *Blade Runner* (Motion Picture). United States: Warner Brothers.

Delmar-Morgan, A. (2013). Toys R Us to stop marketing its toys by gender in wake of sexism claims. *The Independent*. Retrieved from http://www.independent.co.uk/news/business/news/toys-r-us-to-stop-marketing-its-toys-by-gender-in-wake-of-sexism-claims-8798959.html

Devlin, H. (2018). Academic writes 270 Wikipedia pages in a year to get female scientists noticed. *The Guardian*. Retrieved from https://www.theguardian.com/education/2018/jul/24/academic-writes-270-wikipedia-pages-year-female-scientists-noticed

Dionne, E. (2018). The women of Wakanda: Nakia is the real revolutionary of *Black Panther*. *bitchmedia*. Retrieved from https://www.bitchmedia.org/article/the-women-of-wakanda-are-the-real-revolutionaries

Dockterman, E. (2015). Read the *Time Magazine* story that plays a key role in *Steve Jobs*. *Time Magazine*. Retrieved from http://time.com/4063996/steve-jobs-movie-time-magazine-profile/

Doctorow, C. (2014). Feminism and tech: An overdue and welcome manifesto. boingboing.com. Retrieved from http://boingboing.net/2014/05/27/feminism-and-tech-an-overdue.html

Dodson Wade, M. (1994). *Ada Byron Lovelace: The lady and the computer*. New York: Maxwell Macmillan International.

Domonoske, C. (2017). Winner of high school golf tournament denied trophy, because she's a girl. *NPR*. Retrieved from http://www.npr.org/sections/thetwo-way/2017/10/26/560210230/winner-of-high-school-golf-tournament-denied-trophy-because-shes-a-girl

Drew, K., et al. (Producer), & Dean, A. (Director). (2017). *Bombshell: The Hedy Lamarr story* (Motion Picture). United States: Zeitgeist Films.

Dreyfus, B. (2015). Maggie Gyllenhaal just gave a perfect Golden Globes speech. *Mother Jones*. Retrieved from http://www.motherjones.com/mixed-media/2015/01/watch-maggie-gyllenhaal-just-gave-perfect-golden-globes-speech

Ducharme, J. (2018). A woman has finally won the top writing award in comics. *Time Magazine*. Retrieved from http://time.com/5345155/marjorie-liu-eisner-award/

Ellison, M. (Producer), & Jonze, S. (Director). (2013). *Her* (Motion Picture). United States: Warner Brothers.

ENIAC Blog. (2004). The Kathleen McNulty Mauchly Antonelli Story. Retrieved from https://sites.google.com/a/opgate.com/eniac/Home/kay-mcnulty-mauchly-antonelli

England, J., & Cannella, R. (2018). Tweens as technofeminists: Exploring girlhood identity in technology camp. *Girlhood Studies: An Interdisciplinary Journal*, 11, 75–91.

Ensmenger, N. (2010). *The computer boys take over: Computers, programmers, and the politics of technical expertise*. Cambridge, MA: MIT.

Ephron, H. (Producer), & Lang, W. (Director). (1957). *Desk Set* (Motion Picture). United States: 20th-Century Fox.

Erickson, L. (producer and director). (2010). *Top Secret Rosies: The female "computers" of WWII* (Motion Picture). United States: Public Broadcasting Service.

Essinger, J. (2014). *Ada's algorithm: How Lord Byron's daughter Ada Lovelace launched the digital age*. Brooklyn, NY: Melville House.

Estrada, S. (2017). TIME Magazine excluding Tarana Burke from #MeToo cover speaks volumes. DiversityInc.com. Retrieved from https://www.diversityinc.com/news/time-magazine-excluding-tarana-burke-metoo-cover-speaks-volumes

Evans, C. (2018). *Broad band: The untold story of women who made the internet*. New York, NY: Portfolio/Penguin Random House.

Farrow, R. (2017). From aggressive overtures to sexual assault: Harvey Weinstein's accusers tell their stories. *The New Yorker*. Retrieved from https://www.newyorker.com/news/news-desk/from-aggressive-overtures-to-sexual-assault-harvey-weinsteins-accusers-tell-their-stories

Feige, K. (Producer), & Coogler, R. (Director). (2018). *Black Panther* (Motion Picture). United States: Marvel/Walt Disney Studios.

Fenwick & West, LLP. (2016). Gender diversity in Silicon Valley. Retrieved from https://www.fenwick.com/FenwickDocuments/Gender_Diversity_2016.pdf

Frizell, M. (2015). *Female force: Sheryl Sandberg*. Storm Entertainment.

Gay, M. (2000). *Recent advances and issues in computers*. Santa Barbara, CA: ABC/CLIO/Greenwood.

———. (2014). *Bad feminist*. New York: Harper Collins.

———. (2017). *Hunger: A memoir of (my) body*. New York: Harper Collins.

Gigliotti, D., et al. (Producer), & Melfi, T. (Director). (2016). *Hidden Figures* (Motion Picture). United States: 20th-Century Fox.

Bibliography 121

Goldin, C. (2002). A pollution theory of discrimination: Male and female differences in occupations and earnings. National Bureau of Economic Research. Working Paper 8985. Retrieved from http://www.nber.org/papers/w8985

Golemba, B. (1994). Human computers: The women in aeronautical research. Retrieved from https://crgis.ndc.nasa.gov/crgis/images/c/c7/Golemba.pdf

Gould, L. (1978). *X: A fabulous child's story*. New York: Daughters Publishing Company.

Greenblatt, A. (2018). Transgender candidate makes history in a year of "firsts" for women. *Governing: The States and Localities*. Retrieved from http://www.governing.com/topics/politics/gov-vermont-primary-women-midterms-2018-governor-transgender.html

Grossman, L. (2010). Person of the year 2010: Mark Zuckerberg. *Time*. Retrieved from http://content.time.com/time/specials/packages/article/0,28804,2036683_2037183,00.html

Grossman, N., et al. (Producer), & Tyldum, M. (Director). (2014). *The Imitation Game* (Motion Picture). United States: The Weinstein Company.

Grunhut, M., & Machatý, G (Producers), & Machatý, G. (Director). (1933). *Ecstasy* (Motion Picture). Czechoslovakia: Elektafilm.

Hahn, L.D., & Wolters, A.S. (2018). *Women and ideas in engineering: Twelve stories from Illinois*. Urbana, IL: University of Illinois Press.

Hanna, W., & Barbera, J. (Director). (1962). *The Jetsons* (Television Series). United States: Screen Gems.

Haraway, D. J. (1991). *Simians, cyborgs, and women: The reinvention of nature*. New York: Routledge.

Hartman, R. (Producer). (2018). Leading by example to close the gender pay gap. *60 Minutes*. CBS News. Retrieved from https://www.cbsnews.com/news/salesforce-ceo-marc-benioff-leading-by-example-to-close-the-gender-pay-gap/

Harvey, D. (2018). Film review: "Bombshell: The Hedy Lamarr Story." *Variety*. Retrieved from http://variety.com/2017/film/reviews/bombshell-the-hedy-lamarr-story-review-1202621380/

Hicks, B. (2011). The Cherokees vs. Andrew Jackson. *Smithsonian Magazine*. Retrieved from https://www.smithsonianmag.com/history/the-cherokees-vs-andrew-jackson-277394/

Hicks, M. (2017). *Programmed inequality: How Britain discarded women technologists and lost its edge in computing*. Cambridge, MA: MIT Press.

Hickson, A. (2016). THIS is the problem we have with female superheroes. *Refinery29*. Retrieved from http://www.refinery29.com/2016/11/130254/sexist-female-superheroes-comic-book-characters-hypersexual

Hill, C., & Corbett, C. (2010). *Why so few? Women in science, technology, engineering, and mathematics*. Washington, AAUW.

Hill, Collins P., & Bilge, S. (2016). *Intersectionality: Key Concepts*. Cambridge, UK: Polity Press.

Hobson, W., & Boren, C. (2018). Michigan State settles with Larry Nassar victims for $500 Million. *The Washington Post*. Retrieved from https://www.washingtonpost.com/news/early-lead/wp/2018/05/16/michigan-state-settles-larry-nassar-lawsuits-for-500-million/?noredirect=on&utm_term=.a400fbb5ab5b

hooks, b. (1981/2015). *Ain't I a woman: Black women and feminism*. Boston, Mass: South End Press.

Hu, E. (2014). Facebook's diversity numbers are out, and they're what you expect. National Public Radio, All Tech Considered. Retrieved from http://www.npr.org/blogs/alltechconsidered/2014/06/26/325798198/facebooks-diversity-numbers-are-out-and-theyre-what-you-expect

IBM. (n.d.). IBM's ASCC introduction (aka the Harvard Mark I). Retrieved from http://sysrun.haifa.il.ibm.com/ibm/history/exhibits/markI/markI_intro.htm

Iñárritu, A.G., et al. (Producer & Director). (2015). *Birdman* (Motion Picture). United States: Fox Searchlight.

Internet Movie Data Base. (n.d.). Gig Young: Biography. Retrieved from https://www.imdb.com/name/nm0949574/bio

Isaacson, W. (2014). *The innovators: How a group of inventors, hackers, geniuses, and geeks created the digital revolution*. New York: Simon & Schuster.

———. (2007). *Einstein: His life and universe.* New York: Simon and Schuster.
———. (2011). *Steve Jobs.* New York: Simon & Schuster.
Ivins, J. (2004). Retrotype. Retrieved from http://rtetrotype.net.
Jacobs, G. (2015). *The Queen of Code.* (Motion picture). Signals Series ESPN/FiveThirtyEight.
Johnson, S. (2014). *How we got to now: Six innovations that made the modern world.* New York: Riverhead Books/Penguin.
Joy, E. (2015). The other side of diversity. In Shevinksy, E. (Ed.), *Lean out: The struggle for gender equality in tech and start-up culture.* New York, NY: OR\\Books. Kindle Edition.
Judge, M., et al. (Director & Executive Producer). *Silicon Valley.* (Television Series). United States: HBO/Warner Brothers Entertainment.
Kare, S. (2011). Icons. Retrieved from https://kareprints.com
Kamarck, E., Podkul, A.R., & Zeppos, N.W. (2018). The pink wave makes herstory: Women candidates in the 2018 midterm elections. The Brookings Institution. Retrieved from https://www.brookings.edu/blog/fixgov/2018/06/01/the-pink-wave-makes-herstory-women-candidates-in-the-2018-midterm-elections/
Kelion, L. (2015). Steve Wozniak: Shocked and amazed by *Steve Jobs* movie. *BBC News.* Retrieved from http://www.bbc.com/news/technology-34188602
Kennedy, T. R. (1946). Electronic computer figures like a flash. *New York Times.* Retrieved from https://www.nytimes.com/1946/02/15/archives/electronic-computer-flashes-answers-may-speed-engineering-new.html
Kleiman, K., & McMahon, K. (2014). (producers). *The computers: The remarkable story of the ENIAC programmers.* United States: First Byte Productions.
Kramarae, C. (1988). *Technology and women's voices: Keeping in touch.* New York: Routledge & Kegan Paul.
Kubrick, S. (Producer & Director). (1968). *2001: A Space Odyssey.* United Kingdom/United States: Metro-Goldwyn-Mayer.
Kuckler, H. (2017). Silicon Valley upgrades culture for LGBT tech workers. *Financial Times.* Retrieved from https://www.ft.com/content/8dd55500-8efc-11e7-9580-c651950d3672
Kuhn. V. (2017). Remix in the age of Trump. *Journal of Contemporary Rhetoric,* 7(2–3), 87–93.
Lamarr, H. (1966). *Ecstasy and me: My life as a woman.* New York: Bartholomew House.
Lambert, M. (2015). The difference machine: Ada Lovelace, Grace Hopper, and women in tech. *Grantland.* Retrieved from http://grantland.com/hollywood-prospectus/the-difference-machine-ada-lovelace-grace-hopper-and-women-in-tech/
Lazarus, S. (2013). Benedict Cumberbatch and Keira Knightley's *Imitation Game* romance labeled inaccurate. *Radio Times.* Retrieved from http://www.radiotimes.com/news/2013-11-19/benedict-cumberbatch-and-keira-knightleys-imitation-game-romance-labelled-inaccurate
Le Guin, U. (1986). Bryn Mawr commencement address. Retrieved from https://serendip.brynmawr.edu/sci_cult/leguin/
Lessig, L. (2008). *Remix: Making art and culture thrive in the hybrid economy.* New York: Penguin.
Lettice, J. (1998). Corel settles in Lamarr pic lawsuit. *The Register.* Retrieved from http://www.theregister.co.ul/1998/12/02corel_settles_in_lamarr_pic/
Let Toys Be Toys. (2017). Fantastic toys and books to foster STEM skills – Let Toys Be Toys gift guide. Lettoysbetoys.org.uk. Retrieved from http://lettoysbetoys.org.uk/11-fantastic-toys-and-books-to-foster-stem-skills-let-toys-be-toys-gift-guide-2/
Levin, S. (2017). Google told to hand over salary details in gender equality court battle. *The Guardian.* Retrieved from https://www.theguardian.com/technology/2017/jul/17/google-told-to-hand-over-salary-details-in-gender-equality-court-battle
Lewis, T. (2015). Rise of the Fembots: Why artificial intelligence is often female. *LiveScience.* http://www.livescience.com/49882-why-robots-female.html
Lindsay, B. (2017). Read Hillary Clinton's full speech accepting the Wonder Woman award. *Vanity Fair.* Retrieved from https://www.vanityfair.com/style/2017/10/hillary-clinton-wonder-woman-award

Linklater, R. (Producer & Director). (2014). *Boyhood* (Motion Picture), United States: United States: Universal Pictures

Lohr, S. (Apr. 4, 2011). Jean Bartik, software pioneer, dies at 86. *The New York Times*. Retrieved from http://www.nytimes.com/2011/04/08/business/08bartik.html

Lohr, S. (2001). Frances E. Holberton, 84, early computer programmer. *New York Times*. Retrieved from https://www.nytimes.com/2001/12/17/business/frances-e-holberton-84-early-computer-programmer.html

Lorre, C., et al. (Producer & Director). *The Big Bang Theory* (Television Series). United States: Warner Brothers.

Los Angeles Times. (2017). A shift in focus. *The Envelope*. Retrieved from http://www.latimes.com/entertainment/envelope/

Lushington, J., et al. (Producers), & De Emmony, A. (2012). *The Bletchley Circle* (Television Series). United Kingdom: World Productions.

Manian, D., et al. (n.d.) About Feminism. Retrieved from http://aboutfeminism.me

Marenco, S. (2010). *Barbie: I can be a computer engineer*. New York: Random House.

Marshall, F., & Watts, R. (Producers), & Zemeckis, R. (Director). (1988). *Who Framed Roger Rabbit?* (Motion Picture). United States: Touchtone Pictures.

Mattel. (n.d.) Barbie + Tynker Robotics Engineer. Retrieved from https://barbie.mattel.com/en-us/about/barbie-tynker.html

Mattick, J. (2016). Wonder Woman costumes: The evolution of a superheroine. HalloweenCostumes.com. Retrieved from https://www.halloweencostumes.com/blog/p-928-wonder-woman-costumes-the-evolution-of-a-superheroine-infographic.aspx

Maven Youth. (n.d.). Tech Give Back. Retrieved from https://mavenyouth.org/programs/tech-give-back

McAnulty, S. (2017). *Goldie Blox ruins rules the school!* New York: Random House.

McCardle, M. (2017). As a woman in tech I realized: These are not my people. Bloomberg.com. Retrieved from https://www.bloomberg.com/view/articles/2017-08-09/as-a-woman-in-tech-i-realized-these-are-not-my-people

McDonald, S. (2015). Fewer women in top films than there in 2002: Study. *Stuff*. Retrieved http://www.stuff.co.nz/entertainment/film/66048426/fewer-women-in-top-films-than-there-in-2002-study

McGee, A. (2018). Barbie goes abroad: Critiquing feminism, technology, and stereotypes in the narratives and social media strategies of Barbie. In C. Medina and O. Pimentel (Eds.), *Racial shorthand: Coded discrimination contested in social media*. Logan, UT: Computers and Composition Digital Press/Utah State University Press. Retrieved from http://ccdigitalpress.org/book/shorthand/chapter_mcgee.html

McKittrick, S., et al. (Producer), & Peele, J. (Director). (2017). *Get Out* (Motion Picture). United States: Universal Pictures.

Mehta, S. (2010). Meg Whitman reportedly shoved EBay employee in 2007. *Los Angeles Times*. Retrieved from http://articles.latimes.com/2010/jun/15/local/la-me-0615-whitman-20100615

Meledandri, C. (Producer), & Coffin, P., & Balda, K. (Directors). (2015). *Minions* (Motion Picture). United States: Universal.

Merriam-Webster. (2018). Merriam-Webster's 2017 words of the year. Merriam-Webster.com. Retrieved from https://www.merriam-webster.com/words-at-play/word-of-the-year-2017-feminism

Mezrich, B. (2009). *The accidental billionaires: The founding of Facebook, a tale of sex, money, genius, and betrayal*. New York: Doubleday.

Milano, A. (2017). Twitter post. Twitter.com. Retrieved from: https://twitter.com/Alyssa_Milano/status/919659438700670976/photo/1?tfw_site=guardian&ref_src=twsrc%5Etfw&ref_url=https%3A%2F%2Fwww.theguardian.com%2Fculture%2F2017%2Fdec%2F01%2Falyssa-milano-mee-too-sexual-harassment-abuse

Miller, C.C. (2013). Ad takes off online: Less doll, more awl. *New York Times*. Retrieved from http://bits.blogs.nytimes.com/2013/11/20/a-viral-video-encourages-girls-to-become-engineers/?_r=0

Misa, T. (2010). *Gender codes: Why women are leaving computing*. Hoboken, NJ: Wiley.

Misch, G., et al. (Producer & Director). (2004). *Calling Hedy Lamarr* (Motion Picture). Austria/United Kingdom: Mischief Films.

Moeslein, A. (2015). Reese Witherspoon's moving speech at *Glamour*'s 2015 Women of the Year Awards: "Like Elle Woods, I do not like to be underestimated." *Glamour*. Retrieved from https://www.glamour.com/story/reese-witherspoon-women-of-the-year-speech

Molla, R. (2017). A black woman in tech makes $79,000 for every $100,000 a white man makes. Recode.net. Retrieved from https://www.recode.net/2017/4/4/15160924/silicon-valley-women-race-salary-companies-average-less-data-men-tech

Moore, L. (Producer), & Burke, M. (Director). (1999). *Pirates of Silicon Valley*. (Television Film). United States: Turner Network Television.

Morais, B. (15 Oct. 2013). Ada Lovelace, The first tech visionary. *The New Yorker*. Retrieved from http://www.newyorker.com/tech/elements/ada-lovelace-the-first-tech-visionary

Munro, E. (2013). Feminism: A fourth wave? *Political Insight*, 4, 22–25.

Myers, B. (2018). Women and minorities in tech, by the numbers. *Wired*. Retrieved from https://www.wired.com/story/computer-science-graduates-diversity/

Myers, N., et al. (Producer), & Zieff, H. (Director). (1980). *Private Benjamin* (Motion Picture). United States: Warner Brothers.

NASA. (10 August 2018). Mary Ross: A hidden figure. NASA History. Retrieved from https://www.nasa.gov/image-feature/mary-ross-a-hidden-figure

National Public Radio. (2016). "Hidden Figures": How black women did the math that put men on the moon. Retrieved from https://www.npr.org/2016/09/25/495179824/hidden-figures-how-black-women-did-the-math-that-put-men-on-the-moon

——— (2014). After backlash, computer engineer Barbie gets new set of skills. Retrieved from http://www.npr.org/2014/11/22/365968465/after-backlash-computer-engineer-barbie-gets-new-set-of-skills

———. (2015). Grace Hopper, "The Queen of Code," would have hated that title. Retrieved from http://www.npr.org/blogs/alltechconsidered/2015/03/07/390247203/grace-hopper-the-queen-of-code-would-have-hated-that-title

National Women's History Museum. (12 Mar 2018). #OnThisDay. Facebook. Retrieved from https://www.facebook.com/womenshistory/

Nativiad, A. (2017). Meet Molly, the inventive star of GE's latest inspirational ad. *AdWeek*. Retrieved from http://www.adweek.com/creativity/meet-molly-the-inventive-star-of-ges-latest-inspirational-ad/

Nayfack, N. (Producer), & Wilcox, F. (Director). (1956). *Forbidden Planet* (Motion Picture). United States: Metro-Goldwyn-Mayer.

N'Duka, A. (2018). Middleton Media, Google team on biopic about computer scientist Grace Hopper. Deadline Hollywood. Retrieved from http://deadline.com/2018/02/middleton-media-google-grace-hopper-biopic-1202305817/

New York Times. (2017). Jessica Bennett, our new Gender Editor, answers your questions. Retrieved from https://www.nytimes.com/2017/12/13/reader-center/jessica-bennett-our-new-gender-editor-answers-your-questions.html

Noble, S. (2018). *Algorithms of oppression*. New York: New York University Press.

O'Connor, L. (2015). American Studies 7700 Reflection. Unpublished document.

Ohlheiser, A. (2017). The woman behind "Me Too" knew the power of the phrase when she created it—10 years ago. *The Washington Post*. Retrieved from https://www.washingtonpost.com/news/the-intersect/wp/2017/10/19/the-woman-behind-me-too-knew-the-power-of-the-phrase-when-she-created-it-10-years-ago/?utm_term=.bb2c2a542f71

Overlooked. (2018). *New York Times*. Retrieved from https://www.nytimes.com/interactive/2018/obituaries/overlooked.html

Padnani, A. (2018). How an obits project on overlooked women was born. *New York Times*. Retrieved from https://www.nytimes.com/2018/03/08/insider/overlooked-obituary.html

Padua, S. (2015). *The thrilling adventures of Lovelace and Babbage*. New York: Pantheon.

Pao, E. (2017). This is how sexism works in Silicon Valley. My lawsuit failed. Others won't. *The Cut*. Retrieved from https://www.thecut.com/2017/08/ellen-pao-silicon-valley-sexism-reset-excerpt.html

———. (2017). *Reset: My fight for inclusion and lasting change*. New York: Spiegel & Grau.

Pauly. M. (Mar./Apr. 2017). "I Made That Bitch Famous": A brief history of men getting credit for women's accomplishments. *Mother Jones*. Retrieved from: http://www.motherjones.com/media/2017/03/men-taking-credit-women-history

Peck, E. (2016). Marissa Mayer calls out media for sexist coverage. *Huffington Post*. Retrieved from https://www.huffingtonpost.com/entry/marissa-mayer-calls-out-media-for-sexist-coverage_us_57965ea9e4b01180b52fd057

Petit, C., and Sarkeesian, A. (2017). Wonder Woman: The hero we need in a film that falls short. *Feminist Frequency*. Retrieved from https://feministfrequency.com/2017/06/05/wonder-woman-the-hero-we-need-in-a-film-that-falls-short/

Poland, B. (2016). *Haters: Harassment, abuse, and violence online*. Lincoln, NE: University of Nebraska Press.

Project Include. (n.d.). About Project Include. Retrieved from http://projectinclude.org/about

Quinn, Z. (2017). *Crash override: How Gamergate (Nearly) destroyed my life, and how we can win the fight against online hate*. New York, NY: Public Affairs Books.

Racine, E. (2017). #NotInvisible: The Plight of Native American Women and Sexual Violence. Lakota People's Law Project. Retrieved from https://www.lakotalaw.org/news/2017-12-05/notinvisible

Randall, A. (2006). The Eckert tapes: Computer pioneer says ENIAC team couldn't afford to fail—and didn't. *Computerworld*. Retrieved from https://www.computerworld.com/article/2561559/computer-hardware/the-eckert-tapes--computer-pioneer-says-eniac-team-couldn-t-afford-to-fail----and-.html

Recode Staff. (2018). Full video and transcript: Snap CEO Evan Spiegel at Code 2018. Retrieved from https://www.recode.net/2018/5/30/17397120/snap-ceo-evan-spiegel-transcript-code-2018

Recode/MSNBC Staff. (2018). *Revolution: Google and YouTube Changing the World*. Retrieved from https://www.youtube.com/watch?list=PLDIVi-vBsOExzqg6JEfk35vd5AUrLFYf_&v=_M_rSFBYEe8

Reitman, I., Pascal, A. (Producer), & Feig, P. (Director). (2016). *Ghostbusters: Answer the Call* (Motion Picture). United States: Columbia Pictures.

Reinharz, S. (1992). *Feminist methods in social research*. Oxford, UK: Oxford University Press.

Rivera, G. (13 Feb. 2017). (GeraldoRivera). Post. Facebook. Retrieved from https://www.facebook.com/GeraldoRivera/posts/1409432595756423

Rhodes, R. (2011). *Hedy's folly: The life and breakthrough inventions of Hedy Lamarr, the most beautiful woman in the world*. New York: Doubleday.

Robbins, T. (2007). *Hedy Lamarr and a secret communication system*. Mankato, MN: Capstone Press.

Roberts, P. (2013). Why Johnny can't code. Veracode.com. Retrieved from https://www.veracode.com/blog/2013/04/why-johnny-cant-code

Robnett, R. (2016). Gender bias in STEM fields: Variation in prevalence and links to STEM self-concept. *Psychology of Women Quarterly*, 40(1), 65–79.

Romano, N. (13 Jan 2017). Octavia Spencer buys out *Hidden Figures* screening for low income families. *Entertainment Weekly*. Retrieved from http://ew.com/movies/2017/01/13/octavia-spencer-hidden-figures-low-income-families/

Rose, S. (2015). Michael Fassbender on playing Steve Jobs: "Was he flawed? Yeah! We all are." Retrieved from *The Guardian*. https://www.theguardian.com/film/2015/nov/12/steve-jobs-well-done-portraying-the-apple-legend-on-screen

Rosenberg, M., Confessore, N., & Cadwalladr, C. (2018). How Trump consultants exploited the Facebook data of millions. *New York Times*. Retrieved from https://www.nytimes.com/2018/03/17/us/politics/cambridge-analytica-trump-campaign.html?mtrref=www.google.com&gwh=D5AEA269498222575BC89F2461DD1F89&gwt=pay

Rosin, H. (2012). Why doesn't Marissa Mayer care about sexism? *Slate*. Retrieved from http://www.slate.com/blogs/xx_factor/2012/07/16/new_yahoo_ceo_marissa_mayer_does_she_care_about_sexism_.html

Rovan, C., et al. (Producer), & Jenkins, P. (Director). (2017). *Wonder Woman*. United States: Warner Brothers.

Russian, A. (2018). Who was Hedy Lamarr? All about the tragic film star and secret inventor in tonight's *Timeless*. *People Magazine*. Retrieved from http://people.com/tv/who-was-hedy-lamarr-timeless/

Sager, J. (2018). The real reason T. J. Miller left "Silicon Valley." *Page Six*. Retrieved from https://pagesix.com/2018/03/07/the-real-reason-t-j-miller-left-silicon-valley/

Sandberg, S., & Scovell, N. (2013). *Lean in: Women, work, and the will to lead*. New York: Alfred A. Knopf.

Satrun, S. (2013). We are all Wonder Women! *Sarah's Sketchbook*. Retrieved from http://sarahsatrun.blogspot.com/2013/04/we-are-all-wonder-women.html

Scherick, E. (Producer), & Forbes, B. (Director). (1972). *The Stepford Wives* (Motion Picture). United States: Columbia Pictures.

Selber, S. (2004). *Multiliteracies for a digital age*. Carbondale, IL: Southern Illinois University Press.

Sevros, R. (2000). Lamarr, sultry star who reigned in Hollywood of 30s and 40s, dies at 86. *New York Times*. Retrieved from https://www.nytimes.com/2000/01/20/arts/hedy-lamarr-sultry-star-who-reigned-in-hollywood-of-30-s-and-40-s-dies-at-86.html

Sharma, S. (2018). Going to work in mommy's basement. *Boston Review*. Retrieved from http://bostonreview.net/gender-sexuality/sarah-sharma-going-work-mommys-basement

Shetterly, M.L. (2016). *Hidden figures: The story of the African-American women who helped win the space race*. New York: William Morrow and Company.

Sheils, M. (1975). Why Johnny can't write. *Newsweek*, 58–75.

Sheridan, M. (2018). Knotworking collaborations: Fostering community engaged teachers and scholars. In K.L. Blair & L. Nickoson (Eds.), *Composing Feminist Interventions: Activism, Engagement, Praxis*. Fort Collins, CO: The WAC Clearinghouse. Retrieved from https://wac.colostate.edu/books/perspectives/feminist

Shevinsky, E. (2015). *Lean out: The struggle for gender equality in tech and start-up culture*. New York, NY: OR Books.

Siebel Newsom, J, & Costanzo, J. (producer). (2011). *Miss Representation* (Motion Picture). United States: Girls Club Entertainment.

Siede, C. (2015). CTO Megan Smith explains how women in tech are erased from history. *boingboing.com*. Retrieved from http://boingboing.net/2015/05/08/cto-megan-smith-explains-how-w.html

Siltanen, R. (2013). The real story of Apple's "Think Different" campaign. *Branding Strategy Insider*. Retrieved from https://www.brandingstrategyinsider.com/2013/02/the-real-story-of-apples-think-different-campaign.html#.Wl_w7HgSBBU

Sims, D. (2016). The ongoing outcry against the Ghostbusters remake. *The Atlantic*. Retrieved from https://www.theatlantic.com/entertainment/archive/2016/05/the-sexist-outcry-against-the-ghostbusters-remake-gets-louder/483270/

Smith, A. (2016). Please internet responsibly: Rhetorical (techno)feminist models for a digital age. Presentation at the 2016 Thomas Watson Conference on Rhetoric and Composition. Louisville, KY. Retrieved from https://www.slideshare.net/allegrasmith1/please-internet-responsibly-rhetorical-technofeminist-methodologies-for-a-digital-age

Smith, G. (2007). Unsung innovators: Jean Bartik: ENIAC programmer. *ComputerWorld*. Retrieved from https://www.computerworld.com/article/2540042/it-management/unsung-innovators--jean-bartik--eniac-programmer.html

Soderbergh, S., Ekins, S. (Producer), & Ross, G. (Director). 2018. *Ocean's 8* (Motion Picture). United States: Warner Brothers.

Solatoroff, P. (9 Sept. 2015). Trump seriously: On the trail with the GOP's tough guy. *Rolling Stone*. Retrieved from http://www.rollingstone.com/politics/news/trump-seriously-20150909

Bibliography

Solnit, R. (2008). Men explain things to me. TomDispatch.com. Retrieved from http://www.tomdispatch.com/blog/175584/

———. (2014). Poison apples. *Harpers*. Retrieved from https://harpers.org/archive/2014/12/poison-apples/

Stabile, C. (1994). *Feminism and the technological fix*. Manchester, UK: Manchester University Press.

Statistica. (2015). Distribution of computer and video gamers in the United States from 2006 to 2015, by gender. Retrieved from http://www.statista.com/statistics/232383/gender-split-of-us-computer-and-video-gamers

Statt, N. (29 Sept. 2016). Melinda Gates turns her focus to promoting women in tech. *The Verge*. Retrieved from http://www.theverge.com/2016/9/29/13110070/melinda-gates-women-in-tech-microsoft-foundation

Sterling, D. (2017). Dear Google memo writer: The problem's not biology—It's guys like you. *Fortune Magazine*. Retrieved from https://fortune.com/2017/08/10/google-diversity-memo-goldieblox/

Stern, J.M., & Hulme, M. (Producers), & Stern, J.M. (Director). *Jobs*. (Motion Picture). United States: Five Star Feature Films.

Sullivan, G. (2014). Google statistics show Silicon Valley has a diversity problem. *Washington Post*. Retrieved from http://www.washingtonpost.com/news/morning-mix/wp/2014/05/29/most-google-employees-are-white-men-where-are-allthewomen/

Sydell, L. (6 Oct. 2014). The forgotten female programmers who created modern tech. *All Tech Considered*. National Public Radio. Retrieved from: http://www.npr.org/sections/alltechconsidered/2014/10/06/345799830/the-forgotten-female-programmers-who-created-modern-tech

Taub, A. (15 Sept. 2015). What was really going on in that awkward debate moment about putting a woman on the $10. *Vox*. Retrieved from http://www.vox.com/2015/9/17/9347307/gop-debate-woman-ten-dollar

tableflip dot club. (n.d.). @tableflipclub. Retrieved from tableflip.club

The Computation Laboratory of Harvard University. (1946). *A manual of operation for the automatic sequence controlled calculator*. Cambridge, MA: Harvard University Press. Retrieved from https://books.google.com/books?id=atcmAAAAMAAJ&q=bibliogroup:%22Annals+of+the+Computation+Laboratory+of+Harvard+University%22&dq=bibliogroup:%22Annals+of+the+Computation+Laboratory+of+Harvard+University%22&hl=en&sa=X&ved=0ahUKEwjZ1NefrPDcAhUPnFkKHXKzACsQ6AEIKTAA

The Pussyhat Project (n.d.). FAQ. Retrieved from https://www.pussyhatproject.com/faq/

The Representation Project (n.d.). Miss Representation. Retrieved from http://therepresentationproject.org/film/miss-representation/

Tibken, S. (2015). Steve Jobs' legacy includes the women he inspired. CNET.com. Retrieved from https://www.cnet.com/news/steve-jobs-legacy-includes-the-women-he-inspired/

Tiku, N. (2015). Sexism and consequences at TechCrunch's annual award show. *The Verge*. Retrieved from https://www.theverge.com/2015/2/9/8004101/sexism-and-consequences-at-techcrunch-s-annual-award-show

University of Pennsylvania. (n.d.) ENIAC: Celebrating Penn engineering history. Retrieved from http://www.seas.upenn.edu/about-seas/eniac/of-interest.php

U.S. Equal Employment Opportunity Commission. (2016). Diversity in high tech. Retrieved from https://www.eeoc.gov/eeoc/statistics/reports/hightech/

Vagianos, A. (8 Feb. 2017). Ad imagines a world where we treat female scientists like celebrities. *Huffington Post*. Retrieved from http://www.huffingtonpost.com/entry/ad-imagines-a-world-where-we-treat-female-scientists-like-celebrities_us_589b58c1e4b0c1284f29e0c6

Vare, E.A., & Ptacek, G. (1988). *Mothers of invention from the bra to the bomb: Forgotten women & their unforgettable ideas*. New York: Morrow.

Vee, A. (2017). *Coding literacy: How computer programming is changing writing*. Cambridge, MA: MIT Press.

Wajcman, J. (1991). *Feminism confronts technology*. University Park: Penn State Press.

———. (2001). Foreword. In F. Henwood, H. Kennedy, & N. Miller (Eds.), *Cyborg Lives: Women's Technobiographies* (pp. 7–8). York, UK: Raw Nerve Press.

———. (2004). *TechnoFeminism*. Cambridge, UK: Polity Press.

———. (2010). Feminist theories of technology. *Cambridge Journal of Economics*, 34, 143–52.

———. (2015). *Pressed for time: The acceleration of life in digital capitalism*. Chicago: The University of Chicago Press.

Wajcman, J., & Pham Lobb, L.A. (2007). The gender relations of software work in Vietnam. *Gender, Technology, and Development*, 11(1), 1–26.

Wallmark, L., & Chu, A. (2015). *Ada Byron Lovelace and the thinking machine*. Berkeley, CA: Creston Books.

Wallmark, L., & Wu, K. (2017). *Grace Hopper: Queen of computer code*. New York: Sterling Children's Books.

Wanderings.net. (n.d.). Wanderings: Who is Anya Major?—Truth about Apple's 1984 commercial. Retrieved from http://www.wanderings.net/notebook/Main/TruthAboutApples1984Commercial

Warner, J., & Corley, D. (2017). The women's leadership gap. Center for American Progress. Retrieved from https://www.americanprogress.org/issues/women/reports/2017/05/21/432758/womens-leadership-gap/

Watch The Yard. (2018). Barbie's new doll will honor Alpha Kappa Alpha NASA mathematician, Katherine Johnson. Watchtheyard.com. Retrieved from https://www.watchtheyard.com/akas/barbies-new-doll-will-honor-alpha-kappa-alpha-nasa-mathematician-katherine-johnson/

Watters, A. (2015). Men (still) explain technology to me: Gender and education technology. *boundary 2: An International Journal of Literature and Culture*. Retrieved from http://boundary2.org/2015/04/22/men-still-explain-technology-to-me-gender-and-education-technology/

Weingarten, S. (2017). Twitter is dragging this magazine cover for one reason. *College Candy*. Retrieved from https://collegecandy.com/2017/12/22/the-envelope-la-times-cover-diversity-tweets-twitter-photos-details/

Weisbren, B. (Producer), & Leondis, T. (Director). (2017). *The Emoji Movie* (Motion Picture). United States: Sony Pictures.

Weisgram, E.S., Fulcher, M., Dinella, L.M. (2014). Pink gives girls permission: Exploring the roles of explicit gender labels and gendered-type colors on pre-school children's toy preferences. *Journal of Applied Development Psychology*, 35, 401–9.

Wanger, W. (Producer). & Cromwell, J. (Director). (1938). *Algiers*. United States: United Artists.

West, B. & Cohen, J. (Producers and Directors). (2018). *RBG* (Motion Picture). United States: Magnolia Pictures.

White, A. (2017a). Female Marvel Comics editor harassed online for milkshake selfie. *The Telegraph*. Retrieved from http://www.telegraph.co.uk/books/news/female-marvel-comics-editor-harassed-online-milkshake-selfie/

———. (2017b). Marvel fans in uproar after boss says diversity is killing comic sales. *The Telegraph*. http://www.telegraph.co.uk/books/news/marvel-boss-creates-uproar-saying-diversitys-killing-comic-sales/

Wikipedia.org. (24 July 2018). Neuroticism. Retrieved from https://en.wikipedia.org/wiki/Neuroticism

Wilde, J. (2007). *Travelling to infinity: My life with Stephen*. London: Alma Books.

Wilding, F. (1998). Where is the feminism in cyberfeminsm? n. paradoxa, 2, 6–12.

Williams, L. (2015). Actress Felicia Day opens up about GamerGate fears, has her private details exposed minutes later. ThinkProgress.org. Retrieved from http://thinkprogress.org/culture/2014/10/23/3583347/felicia-day-gamergate

Witherspoon, R., et al. (Producer), & Vallee, J. (Director). (2014). *Wild* (Motion Picture). United States: Fox Searchlight.

Wong, M., & Svrlug, S. (11 Feb. 2017). Yale renames Calhoun College because of historic ties to white supremacy and slavery. *The Washington Post*. Retrieved from https://

www.washingtonpost.com/news/grade-point/wp/2017/02/11/yale-renames-calhoun-college-because-of-historic-ties-to-white-supremacy-and-slavery/?utm_term=.ef49c6dbd93d
Wood, B. (2014). The history of WiFi. Purple.ai. Retrieved from https://purple.ai/blogs/history-wifi/
Yang, G. L., & Holmes, M. (2015). *Secret coders*. New York: Humble Comics.
———. (2016). *Secret coders: Paths & portals*. New York: Humble Comics.
———. (2017). *Secret coders: Secrets and consequences*. New York: Humble Comics.
———. (2017). *Secret coders*: *Robots and repeats*. New York: Humble Comics.
Yiannopoulos, M. (2016). Full text: Milo on how feminism hurts men and women. *Breitbart*. Retrieved from https://www.breitbart.com/milo/2016/10/07/full-text-milo-feminism-auburn
Zimbalist, S. (Producer). & Conway, J. (Director). (1940). *Boom town*. United States: Metro-Goldwyn-Mayer.
Zimmerman, J. (2015). Screw leaning in. It's time to slam the door in Silicon Valley's face. *The Guardian*. Retrieved from http://www.theguardian.com/commentisfree/2015/apr/15/screw-leaning-in-its-time-to-slam-the-door-in-silicon-valleys-face

Index

4Chan, 81
60 Minutes, 37

AAUW, 12, 35, 106, 109
Aberdeen Proving Ground, 33
About Feminism, 10
Abrams, Stacey, xv
Ada Developer Academy, 18
Ada Lovelace Day, 18
Ada: A Journal of Gender, New Media, and Technology, 18
Aiken, Howard, 25, 26, 37, 39, 40, 44
Algiers, 19
Amanpour, Christianne, 60
American Association of University Women. *See* AAUW
American Born Chinese, 88
Analytical Engine, 17
Anita Borg Institute, 49
Antheil, George, 17, 19–22, 25, 38, 46, 57, 89, 100, 103, 112, 114
Antonelli, Kathleen McNulty Mauchly, 25, 29, 33, 35, 43, 91, 101
Antos, Heather, 82, 85
Apple Computer, 50, 53, 55, 58
Apple II, 58
Arquette, Patricia, 8, 48
Artificial Intelligence, 13, 14
Atari, 54
Automatic Control Sequence Calculator. *See* Mark I

Babbage, Charles, 16–18, 19–20, 22, 25, 39–40, 89, 92, 100, 112
Bader Ginsburg, Ruth, xvi
Bialik, Mayim, 11
Ballet Mécanique, 21
Barbie, 5
Barbie: I Can Be a Computer Engineer, 105
Barnes, Susan, 55
Bartik, Jean Jennings, 25, 29, 34, 35–36, 41, 43, 92, 96, 97, 101
Beevis and Butt-head, 67
Benioff, Mark, 71
Berniebros, 81
The Big Bang Theory, 7
Bill and Melinda Gates Foundation, 59, 64
BINAC, 41
Birdman, 1
Black Girls Code, 11, 69
Black Panther, 82, 112
Black Women in Computing, 49
BlackComputeHER, 49
Blade Runner, 56
Bletchley Circle, 7
Bletchley Park, 29
Bloch, Richard, 39
Bok, Mary Louise, 19
Bombshell: The Hedy Lamarr Story, 72, 114
Bono, 59
Boom Town, 19

Borg, Anita, 45, 49
Boyer, Charles, 19
Boyhood, 8
Boyle, Danny, 59
Breitbart, 3
Brennan, Chrisann, 54
Brown, James Cooke, 75
Buffet, Warren, 64
Bullock, Sandra, 106
Bündchen, Gisele, 14
Bureau of Labor Statistics, 8
Burke, Tarana, 102
Butler, Judith, x
Byron, Lord, 15

Calandrelli, Emily, 92
Calhoun College, 41, 46
Calhoun, John C., 41
Calling Hedy Lamarr, 18, 72
Cambridge Analytica, 104, 110
Careers boardgame, 77
Carly for America, 28
Carson, Ben, 29
Center for American Progress, ix
Chan Zuckerberg Initiative, 64
Chan, Priscilla, 63
Charles Babbage Institute of Science and Engineering, 18
Charles Babbage Research Center, 18
ChickTech, 109
Clarke, Arthur C., 57
Clarke, Joan, 2, 45, 87
class, x, xi, xii, xiii, xvii, 8, 9, 10, 16, 17, 26, 28, 38, 45, 54, 71, 84, 87, 100
Clinton, Hillary, xvi, 3, 27, 81, 86, 87
Coates, Ta-Nehisi, 82–83
code, 1, 6, 7, 12, 30, 37, 39, 41, 45, 48, 63, 68, 80, 89, 90, 91, 107, 110, 113
Code Conference, 70
codebabes.com, 12, 110
codedicks.com, 12
collaboration, xx, 7, 15, 20, 22, 25, 26, 30, 31, 33, 34, 36, 37, 38, 40, 41, 43, 44, 45, 53, 54, 55, 58, 59, 60, 63–64, 68, 72, 73, 82, 89, 90, 95, 100, 103, 108, 112
COLOSSUS, 7
Comic Con International, 82
Computer History Museum, 12, 31, 72

The Computers, 35
Cook, Tim, xiv, 57, 58
cosplay, 82, 84
Crazy Ones campaign, 57
Crenshaw, Kimberlé, xi
Croft, Lara, 82
Cromar, Stewart Lamb, 96
Cunningham, Andy, 55
Curtis School of Music, 19
Czarnecki, Kazimierz, 46

Damore, James, 67, 79, 80, 87, 88
Davis, Juliet, 105
De Morgan, Augustus, 16
Depression Quest, 3
Dern, Laura, 8
Desk Set, xx, 26, 31–33, 36
Difference Engine, 16, 39, 40
Digital Media Academy, 109
Digital Mirror Computer Camp, 106, 108
Dorsey, Jack, 60
Dresselhaus, Mildred Millie, 48, 73, 93

Eckert-Mauchly Computer Corporation, 41
Eckert, J. Presper, 25, 34, 37, 44
Ecstasy, 16
Ecstasy and Me: My Life as a Woman, 20
Einstein, Albert, 53
Einstein: His Life and Universe, 99
Eisner Awards, 83
Electronic Dream Phone, 77
Electronic Frontier Foundation, 15, 73
Electronic Numerical Integrator and Computer. *See* ENIAC
Emoji Movie, 94
empowerment, ix, xii, 5, 12, 18, 71, 86, 87, 91, 96, 109, 109–110, 115
ENIAC, xx, 7, 31, 45, 50, 81
Equal Opportunity Commission, ix

Facebook, 8, 35, 45, 50, 110
Farrow, Ronan, 101
feminism, x–xv, 3, 5–7, 10, 18, 26, 28, 37, 60, 65, 66, 69, 71, 76, 78, 80, 86, 100, 101, 103, 106, 111, 113
Feminist Frequency, 86
Field, Sally, 87
Fiorina, Carly, 27
Forbidden Planet, 13

Franken, Al, 101

Gable, Clark, 19
Gaga, Lady, 45
#gamergate, 3, 4, 62, 81
Gates, Bill, 3, 6, 50, 55, 59, 59–60, 62, 64, 71, 81, 104, 113
Gates, Melinda, 50, 51
Gay, Roxane, 8, 82
General Electric, 48, 93
Genius, 99
genius narrative, xviii, xx, 4, 7, 16, 22, 31, 36, 37, 39, 44, 45, 51, 53, 54, 57–58, 59, 61, 64, 72, 80, 88–89, 99
George Washington University, 68
Get a Mac Campaign, 93
Get Out, 111
Ghostbusters, 3, 106
Girl Talk, 77
Girl Talk Date Line, 77
Girls Who Code, 10
Glamour Magazine Woman of the Year, 87
Golda Ross, Mary, xvi
GoldieBlox, 11
Golemba, Beverly, 47, 48
Google, xvi, xviii, 8, 35, 61, 67, 79, 115
Gould, Lois, 75, 91
GQ Magazine, 57
Grace Hopper Celebration of Women in Computing, 49
Grand Theft Auto, 82
Greywind, Savanna LaFountaine, xx
Gyllenhaal, Maggie, 8, 48

Hack the Hood, 69
Hall, Anthony Michael, 59
Hallquist, Christine, xv
Hamilton, Margaret, 48, 88, 95
Harvard Computational Laboratory, 39
HarvardConnection, 63
hashtag activism, xiii, 82
Hawking, Stephen, 1, 22, 72, 94
Hawn, Goldie, 87
HBO, 68
Heitkamp, Heidi, xviii
Hepburn, Katharine, xx, 26
Her, 13–14
Hewlett Packard, 27, 60

Hidden Figures, 8, 45, 48
Hill Collins, Patricia, x
historiography, xiii, 8, 23, 96, 112
Hoffman, Joanna, 55
Holberton, Frances Betty Snyder, 25, 28, 35, 43, 101
The Honourable Woman, 8
hooks, bell, xii–xiii
Hopper, Grace, 6, 8, 36, 37, 40, 41, 59, 60, 72, 81, 88, 106
Hughes, Chris, 63
Human Computer Project, 45

IBM, 25, 39
The Imitation Game, 1, 7, 12, 26, 27, 35, 87
innovation, x, xiv, xvi, xviii, 4, 5, 7, 8, 11, 16, 17, 22, 23, 25, 26, 30, 37, 39, 40, 42, 44, 45, 49, 50, 53, 54, 55, 56, 57, 59, 64, 73, 78, 82, 88, 89, 94, 100, 103, 110, 111, 115
intersectional, x, xi–xii, xiv–xv, 5, 10, 60, 62, 66, 84, 107, 112
Iowa State University, 47
The Iron Lady, 60–61
Isaacson, Walter, 1, 99

Jackson, Mary, xx, 48, 49, 101, 106
Jackson, Shirley Ann, 8
Jemison, Mae, 95
The Jetsons, 13
Jobs, xx, 7, 53
Jobs, Lisa Brennan, 54
Jobs, Steve, 6, 21, 22, 26, 26–36, 44, 50–51, 53, 54–59, 60, 61, 63, 66, 67–68, 72, 82, 94, 100, 104, 113; relationship with daughter, 54; return to Apple, 56, 57
Johannson, Scarlett, 13
Johnson, Katherine, 45, 48, 49, 88, 94, 95, 101, 106
Jones, Felicity, xvi
Jones, Leslie, 3
Jones, Rashida, 63
Joy, Erica, 69, 70
Judge, Mike, 67

Kare, Susan, 55
Kasich, John, 28

Khan, Kamala, 82
King of the Hill, 67
Kleiman, Kathy, 35, 92
Knightley, Keira, 2
Kutcher, Ashton, 7, 53

labor, x, xi, xiv, xvii–xviii, 5, 8, 11, 14, 16, 30, 35, 46, 61, 65, 66, 79, 87, 94
Lamarr, Hedy, xviii, 15–22, 72, 94, 100, 114
Langley Research Center, 45
Lauer, Matt, 101
Lean In: Women, Work, and the Will to Lead, 8, 28, 65, 71
LEGO Women of NASA, 95
Le Guin, Ursula, 99, 105, 111
Lesbians Who Tech, xiv, 70
Lessig, Lawrence, 83
"Let Books Be Books," 78
"Let Toys Be Toys," 78, 106
LGBTQ, xiv, xvii, 105, 110
Library of Congress National Ambassador for Young People's Literature, 89
literacy, technological, xiii, 12, 15, 76, 89, 90, 91, 106, 107
Liu, Marjorie, 82
Loder, Anthony, 18
Loder, Denise, 18
Loder, John, 18
Logo Computer Language, 90
Lovelace, Ada, xiv, xviii–xx, 1, 4, 7, 15–22, 25–27, 31, 37, 39–40, 43, 45–46, 72, 88, 90, 92–94, 96–97, 100, 106, 112, 113, 114
Lubetich, Shannon, 69
Lucy, 13

MacArthur Fellowship, 88
Macintosh 1984 campaign, 55
Major, Anya, 56
Mall Madness, 77
Mandl, Fritz, 19
Mark I, 6, 39, 40, 90
Martinez, Alitha E., 82
Marvel Comics, 82
Marvel, Ms., 82
Massachusetts Institute of Technology, 8
Massachusetts Interscholastic Athletic Association, 95

Mattel, 95
Mauchly, John, 25, 35, 44
Mayer, Marissa, 3, 60–61, 69, 73
McCollum, Andrew, 63
Melber, Ari, 66
Meltzer, Marlyn Westcoff, 25, 29, 35, 43, 91, 101
Menabrea, Luigi, 17
Merriam-Webster, 100
#MeToo, xiii, 71, 101–102
Microsoft, 55, 59, 93
Milano, Alyssa, 101, 102
Miller, TJ, 68, 69
Minions, 106
misogyny, xv, xvi, xvii, 10, 28, 56, 68, 69, 71, 72, 78
Monstress, 82
Moore, Julianne, 8
Moore, Roy, 101
Moskovitz, Dustin, 63
Mother Theresa, 28
Musk, Elon, 101

Nadella, Satya, 2, 62
NASA, xviii, 8, 46, 47
Nash, Emily, 95
Nassar, Larry, 101
National Advisory Committee for Aeronautics, 45
National Medal of Science in Engineering, 48
National Women's History Museum, 111
Navy Women's Reserve Act, 39
New York University, 38
Newman, Omarosa Manigault, 28
Noble, Safiya, 50, 100
Norma Rae, 87
Not Your Mama's Gamer, 103

O'Connor, Sandra Day, xvi
O'Reilly, Bill, 101
Obama, Barack, 2
Oceans 8, 106
oral history, 37, 38
Orwell, George, 55
OutinTech, 110

Pao, Ellen, 11, 62
Patterson, Neva, 32

Index

Pearl Harbor, 39
Peele, Jordan, 111–112
Pichai, Sundar, 66, 67, 71
Pirates of Silicon Valley, 59
Pollution theory of discrimination, 80
power, x–xi, xiii, xiv, xvii, 5, 26, 28, 49, 59, 65, 68, 69, 71, 88, 91, 108, 109, 110, 111, 112
Presidential Medal of Freedom, 48
Private Benjamin, 87
Pussyhat Project, xvi

Quinn, Zoe, 3, 103

Rabbit, Jessica, 82–83
race, xi, xvii, 9, 10, 11, 49, 50–51, 71, 82, 100, 112, 113
Reddit, 62, 81
Redmayne, Eddie, 1
Reed College, 53
Reinharz, Shumalit, xii, xiii
Rensselaer Polytechnic Institute, 8
Retrotype, 85
Richards, Adria, 4
Ride, Sally, 95
Rivera, Geraldo, 42
Roman, Nancy Grace, 95
Ross, Chief John, xviii

Salesforce, 71–72
San Diego State University, 7
Sandberg, Sheryl, 2–3, 9, 28, 61, 64, 65, 102
Sandberg, Sheryl, 28
Sanders, Bernie, 81
Sarandan, Susan, 72
Sarkeesian, Anita, 4, 86, 87, 103
Savannah's Act, xviii
Saverin, Eduardo, 63
Scott, Ridley, 56
Secret Coders series, 88, 89–90
Sessions, Jeff, xvi
Sharma, Sarah, xvii
Shetterly, Margot Lee, 8
Silent Sentinels, 111
Silicon Valley, xx, 67–69
Silicon Valley, xv, xvii, 60, 63, 67; and racial diversity, 69; as meritocracy vs. mirrortocracy, 66

Silicon Valley Conference for women, 3
The Simpsons, 82
Siri, 14
The Social Network, xx, 55, 62, 63
Society of Women Engineers, 11
Solnit, Rebecca, 23, 55
Sorkin, Aaron, 58, 63
Spence, Frances Bilas, 25, 35, 43, 91, 92, 101
Spencer, Octavia, 48
Spiegel, Evan, 70
STEM, ix, x, xv, xviii, xx, 3, 4, 5, 11, 12, 26, 30, 45, 48, 49–50, 64, 65, 76, 78, 79, 92, 93, 106, 109, 114, 115
The Stepford Wives, 13
Sterling, Debbie, 11, 80
Steve Jobs, 55, 63, 99
Still Alice, 8
storiography, xi, 45, 58, 69, 70, 85, 91, 92, 94
Stravinsky, Igor, 19
Super Bowl, 55
Swift, Taylor, 45
Swisher, Kara, 66
Systers, 49

table flip dot club, 10
Tech Ready Women, 70
Techbridge, 11
tech broculture, xiv, xvii, 3, 46, 69, 80, 81, 97
TechCrunch, 68, 69, 70
technofeminism, xi; activism, xi, 108, 109; counter-narrative, 22, 69, 91; lived experience, 23, 39; material conditions, 49, 53, 78; pedagogies, 83; reliance on narrative, xii; recovery of women's contributions, 7; resistance, 71
Teitelbaum, Ruth Lichterman, 25, 35, 43, 91, 101
Thatcher, Margaret, 60–61
The Theory of Everything, 1, 7, 26, 27, 35
Think Different Campaign, 56
Titstare, 68
Top Secret Rosies, 35
Toys "R" Us, 75, 78
Tracy, Spencer, xx, 19, 26
Trump, Donald, xv, 27–28, 45, 111
Turing, Alan, 1, 22, 72, 87, 94

Twitter, 3, 11, 29, 82, 101, 111, 115

Uber, 67, 68
United Nations, 86
United States Department of Labor Department, 79
UNIVAC, 26, 41
University of California Berkeley, 88
University of Maryland, 115
University of Pennsylvania, 29, 30
University of Virginia, 46

Van Britton Brown, Marie, 8
Vannevar Bush Award, 8
Vassar, 38, 41
Vaughn, Dorothy, 45, 101
VTech, 105

Wajcman, Judy, x, xii, xiii, xviii, 26, 33, 50, 51, 100
Wallmark, Laurie, 92
Weinstein, Harvey, 101
West, Kanye, 45
Whitman, Meg, 3, 60, 64
Who Framed Roger Rabbit?, 82
"Why Johnny Can't Write," 107
Wild, 8
Wilde, Jane, xvi, 1
Wilkerson, Brenda Darden, 49
Williams, Riri, 82
Winklevoss, Cameron and Tyler, 65

Winslet, Kate, 55
Witherspoon, Reese, 8, 87
Wojcicki, Susan, 66, 67
Women and Technology Hall of Fame, 35
Women of ENIAC, 25, 30, 45, 55, 92, 104
Women of Apple, 55
Women's March on Washington, xv
Women's Media Center Wonder Woman Award, 86
Wonder Woman, 86, 112
Wonder Woman, 87, 111
Wozniak, Steve, 53, 58, 68
Wu, Brianna, 11
Wyle, Noah, 59
Wynette, Tammy, 2

"X: A Fabulous Child's Story," 75
Xerox Parc, 59
Xploration Outer Space, 92

Yahoo, 8, 61
Yale University, 41
Yang, Gene Luen, 88, 105
Yiannopoulos, Milos, 3
Young, Gig, 31–32

Zuckerberg, Mark, 3, 50, 59–60, 62–63, 66, 72, 104, 110, 113

About the Author

Kristine L. Blair is professor of English and dean of the College of Liberal Arts and Social Sciences at Youngstown State University. In addition to her publications in the areas of gender and technology, online learning, and graduate student and faculty professional development, Blair currently serves as editor of both the international print journal *Computers and Composition* and its separate companion journal *Computers and Composition Online*. Among her co-edited collections are *Composing Feminist Interventions: Activism, Engagement, Praxis*; *Webbing Cyberfeminist Practice: Communities, Pedagogies, and Social Action*; and *Feminist Cyberscapes: Mapping Gendered Academic Spaces*. Her awards include recognition as a 2017 Distinguished Woman Scholar by her doctoral alma mater, Purdue University, and the 2017 Lisa Ede Mentoring Award from the Coalition of Feminist Scholars in the History of Rhetoric and Composition.